Anticipating Flashpoints with Russia

Patterns and Drivers

SAMUEL CHARAP, SEAN M. ZEIGLER, IRINA A. CHINDEA,
MOLLY DUNIGAN, ALYSSA DEMUS, JOHN J. DRENNAN,
WALTER F. LANDGRAF III, JONATHAN WELCH, GRANT JOHNSON,
GREGORY WEIDER FAUERBACH, NATHAN VEST, MELISSA SHOSTAK

Prepared for the United States Army
Approved for public release; distribution unlimited

RAND ARROYO CENTER

For more information on this publication, visit **www.rand.org/t/RRA791-1**.

About RAND

The RAND Corporation is a research organization that develops solutions to public policy challenges to help make communities throughout the world safer and more secure, healthier and more prosperous. RAND is nonprofit, nonpartisan, and committed to the public interest. To learn more about RAND, visit www.rand.org.

Research Integrity

Our mission to help improve policy and decisionmaking through research and analysis is enabled through our core values of quality and objectivity and our unwavering commitment to the highest level of integrity and ethical behavior. To help ensure our research and analysis are rigorous, objective, and nonpartisan, we subject our research publications to a robust and exacting quality-assurance process; avoid both the appearance and reality of financial and other conflicts of interest through staff training, project screening, and a policy of mandatory disclosure; and pursue transparency in our research engagements through our commitment to the open publication of our research findings and recommendations, disclosure of the source of funding of published research, and policies to ensure intellectual independence. For more information, visit www.rand.org/about/principles.

RAND's publications do not necessarily reflect the opinions of its research clients and sponsors.

Cover: A Russian military convoy travels on a road near Zugdidi, a city in western Georgia, on August 19, 2008; REUTERS / Alamy Stock Photo.

About This Report

The research reported here was completed in May 2021.

This report documents research and analysis conducted in 2019–2021 as part of a project entitled *Anticipating Flashpoints with Russia*, sponsored by U.S. Army Europe. The purpose of the project was to identify possible Russian conflict scenarios in and near the U.S. Army Europe area of responsibility that could entangle the United States, assess these scenarios by level of risk, and evaluate their potential future trajectories. The report was completed before the full-scale Russian invasion of Ukraine that began in February 2022 and has not been subsequently revised.

This research was conducted within RAND Arroyo Center's Strategy, Doctrine, and Resources Program. RAND Arroyo Center, part of the RAND Corporation, is a federally funded research and development center (FFRDC) sponsored by the United States Army. The research reported here was completed in May 2021, followed by security review by the sponsor and the U.S. Army Office of the Chief of Public Affairs, with final sign-off in May 2023.

RAND operates under a "Federal-Wide Assurance" (FWA00003425) and complies with the *Code of Federal Regulations for the Protection of Human Subjects Under United States Law* (45 CFR 46), also known as "the Common Rule," as well as with the implementation guidance set forth in DoD Instruction 3216.02. As applicable, this compliance includes reviews and approvals by RAND's Institutional Review Board (the Human Subjects Protection Committee) and by the U.S. Army. The views of sources utilized in this study are solely their own and do not represent the official policy or position of DoD or the U.S. Government.

Acknowledgments

We would like to thank U.S. Army Europe-Africa (USAREUR-AF) for sponsoring this project, and then-USAREUR-AF Commander, General Christopher G. Cavoli, for his support in this effort. The inputs and feedback that the sponsor provided during the course of this project were hugely helpful. We also thank the sponsor for affording us the latitude to advance our research along several related paths.

We would also like to thank two sets of colleagues at the RAND Corporation. The first is the management team within RAND Arroyo Center broadly and the Strategy, Doctrine, and Resources program specifically. Among those most intimately involved in managing this project and improving the quality of it were the then-program and associate program directors: Jennifer Kavanagh and Stephen Watts. A second set of colleagues greatly contributed to the administration of the project. Of particular note is the contribution of Silas Dustin, whose patience and diligence in formatting and preparing this manuscript cannot be overstated.

Finally, we owe a debt of gratitude to Andy Stravers (RAND) and Keith Darden (American University), whose reviews of this report challenged our thinking in positive ways and greatly improved the final product.

Summary

The research reported here was completed in May 2021.

Even before its full-scale invasion of Ukraine in 2022, Russia had a large number of ongoing and potential disputes with other countries, motivated by a variety of territorial, political, and economic issues. Furthermore, as Moscow has sought to expand its international role, it has increased Russian involvement in civil conflicts, using both overt and covert means. Russian activity in Ukraine, Syria, and Libya raised the prospect that the United States might find itself militarily entangled with Russia in various global hotspots. The objective of this report is to identify possible Russian conflict scenarios in and near the U.S. Army Europe area of responsibility (AOR) that could entangle the United States and present distinct military challenges to the Army. While the research was concluded before Russia's full-scale invasion of Ukraine in 2022, the insights presented here are no less relevant for understanding patterns in Moscow's assertive behavior.

We used a variety of analytical tools to help us better understand and anticipate flashpoints in the Russian context,[1] adopting different approaches (qualitative case studies, quantitative modeling, and an examination of additional Russian activities overseas that could prove escalatory) to examine broader Russian conflict and dispute trends and to derive possible future flashpoints to which the U.S. military—and the Army, in particular—might be called on to respond. In this report, we focus, in particular, on potential flashpoints with non-NATO (North Atlantic Treaty Organization) allies.

Our qualitative examination began by categorizing the nature of past Russian run-ins with other states in the post-Soviet period. For this purpose we distinguished between *flashpoints*, or *militarized conflicts* (defined as

---

[1] We use the term *flashpoint* as a synonym for *militarized conflict*, which we define as Russia's use of force toward another state (or vice versa) that results in one or more combat deaths, or the overt and explicit military intervention by Russia into an intrastate conflict in support of the recognized government resulting in one or more battle deaths. We consider Russian intervention into an intrastate conflict on the side of rebels as an *interstate conflict*.

interstate clashes or interventions in civil war that result in combat deaths) and *militarized disputes* (interstate frictions that did not lead to recorded battle deaths). Using the political science literature on escalation and the causes of conflict, we developed a 16-factor framework of possible drivers. We then used this framework to conduct in-depth case studies of four flashpoints: the Ukraine intervention (2014), the Russia-Georgia War (2008), the Tajik Civil War (1992–1997), and the intervention in the Syrian Civil War (ongoing since 2015). We also examined Russian disputes with Ukraine and Georgia that did not escalate to militarized conflict to understand why they remained below the flashpoint level.

For this report, we also brought to bear quantitative analytical techniques to assess the factors or correlates most associated with the onset of interstate frictions with Russia. This mode of analysis allowed us to also undertake a predictive exercise in an effort to gauge which countries, based on their characteristics, might be most prone to future militarized interstate disputes (MIDs) with Russia.

In addition to these lines of effort (which focused largely on traditional overt interstate clashes or open interventions in civil wars), we examined additional potential drivers of conflict not captured in the historical data. Specifically, we looked at Russia's use of private military contractors (PMCs) and operations in the information environment to see whether and how either might lead to a flashpoint in the future.

Findings

Both our qualitative and quantitative analyses reinforce the centrality of geographic proximity in driving Russian conflicts and disputes. Except the two related to the Syria operation, all Russian militarized conflicts in the time frame of the analysis (1992–2019) have been located in the former Soviet region. The quantitative analysis underscores the correlation of territorial contiguity and former Soviet republic status with MID onset. The case studies further clarify the causal relationship between proximity to Russia and escalation.

That Russia has maintained a willingness to engage militarily in its so-called near abroad is not a revolutionary finding. However, it remains an important fact when considering whether and where Russia might become involved in militarized conflicts or disputes in the future. Russia shares a land border with 16 different countries, more than any other country in the world. This fact, coupled with the correlation of territorial contiguity and MID onset, suggests that the potential for Russian conflict spans many countries and several important regions of the world.

In addition to geographic proximity, our quantitative analysis points to at least two other factors correlated with Russian MIDs. First, states with unregulated land or maritime border issues with Russia are more likely to find themselves involved in a MID with Russia. Historically, disagreements related to territory and the demarcation of national boundaries have often been the source of militarized disputes and engagements. Russia is no exception. As already specified, Russia borders more countries than any other. By our accounting, Russia has ongoing territorial disagreements with six other states, but that number has been as high as 15 since 1992. Where there is ambiguity, uncertainty, or historical precedence for disagreement over borders, there is the potential for a clash.

Another factor to emerge from the quantitative models associated with Russian MID onset with other states is time. Simply put, a recent MID between Russia and another state strongly suggests that another MID might be in the offing. Russian engagement with Georgia—there were multiple disputes between the two in the years leading up to the 2008 war—is an apt example. This finding also suggests that the maxim "time heals all wounds" does apply to Russia's relations with other states.

The data also suggest that Russia has a history of MIDs with both major powers and smaller countries. Looking forward, major powers might expect to find themselves entangled in disputes with Russia. How such engagements are handled will likely dictate their severity; our analysis does not necessarily presage escalation to the point of combat deaths. But the data do suggest that Russia has not shied away from MIDs with major powers in the past.

Our qualitative analysis highlights the role of broader geopolitical factors in driving conflict escalation with Russia. Russia's dissatisfaction with its place in the international system, acute uncertainty about the future, and reputational costs were central to the escalatory dynamic in all four conflicts examined in the case studies of militarized conflict. This was true for both interstate clashes and interventions in civil conflicts. However, external threats to Russia were often the immediate trigger. External security threats were identified as drivers of escalation in all of the case studies. Taken together, these observations reinforce the view of Russia as a status-seeking, geopolitically minded but predominately regional power, or at least a power that sees its immediate environs as the primary source of security threats. Furthermore, the centrality of external threats across the cases suggests an actor driven to war by the perception of imminent potential losses, not by an expansionist instinct—at least before 2022.

Applying the same framework to test for the presence of these potential drivers in cases of disputes with Ukraine and Georgia that did not escalate to militarized conflict reinforced the importance of two of the geopolitical factors: dissatisfaction with the international systemic status quo and acute uncertainty about the future. It also underscored the importance of external threats. These three factors were absent when disputes with the same countries over broadly similar issues did not escalate to the point of a militarized conflict but were present when they did.

Two additional results emerge from the qualitative analysis. Both opposing states in the interstate militarized conflicts we examined were considered democratic at the time, but in neither case did regime type play a causal role in driving escalation. This finding calls into question the commonly encountered notion that Russia is driven to war to prevent democracy from emerging on its periphery. Second, all nine identified militarized conflicts took place in states where Russia had a military installation. We had assumed that the presence of a Russian military facility in the state where a conflict was located could drive escalation by facilitating Moscow's intervention. However, in those cases where the presence of a facility was an important driver of escalation, these facilities served as liabilities—deepening Russia's involvement in the conflict because its forces came under threat—rather than as assets that allowed for enhanced power projection beyond Russian borders. Even Russia's alliance obligations only played into *disputes* when its own forces come under threat.

Finally, our examination of two additional potential drivers of conflict—operations in the information environment (OIE) and the use of PMCs—reveals several important dynamics. A first finding is somewhat counter-intuitive: There are no cases of OIE that led to an interstate conflict, and even the most plausibly escalatory of these operations have not come close to producing a flashpoint. We document two such operations, a cyberattack on a Saudi oil refinery and an electromagnetic attack on military and civilian assets during a NATO exercise in Northern Europe, that arguably came the closest to escalating. These examples demonstrate that such OIE can have significant strategic effects and thus merit continued attention in the future. The analysis suggests that the characteristics of OIE most likely to escalate to a flashpoint are (1) an attributed attack that (2) inflicts physical damage.

Russian use of PMCs in zones of conflict or unstable countries has the potential to ignite an interstate clash. Certainly, the 2018 Deir Ezzor incident in Syria came precipitously close to producing such an outcome. The tactical engagement of PMCs in a paramilitary role and their targeting of a capable adversary or partner force appear to be the PMC-related variables most strongly correlated with escalatory dynamics.

Implications

The implications of this report for the Army do not follow in a straightforward manner from the research findings. We examined Russia's pre-2020 history of conflicts and disputes and what that history might portend for the future; we do not prescribe U.S. responses to potential future scenarios. That said, there are a variety of implications for the Army that emerge from the research presented here. First, the Army might be called on to conduct or complement operations to counter Russian flashpoints with countries that, particularly before 2022, were not central to Army planning considerations.

A second implication stemming from our findings concerns the frequency and propensity for Russia to engage in both militarized conflicts and disputes. Militarized conflicts involving Russia were relatively rare and have not involved major powers to date. Militarized disputes, however, are not infrequent. Although our data set on disputes ends in 2010, anecdotal evidence suggests that disputes might have been even more common in recent years.

A third key implication stems from our finding about the centrality of proximity in driving escalation: Army engagements with countries in Russia's immediate periphery in post-Soviet Eurasia could be a source of a dispute or even conflict involving Russia. Appreciating this potential in advance can help mitigate future incidents and possibly prevent escalation.

Finally, such Russian activities as the deployment of PMCs and use of OIE are problematic but have not produced interstate conflict to date. Although the Deir Ezzor incident indicates that PMC tactical operations can lead to U.S. military engagements with these parastatal actors, even that case did not become an interstate clash. Moreover, it is not obvious that the United States, and the Army in particular, should respond in kind to every Russian provocation involving OIE or use of PMCs. Publicly highlighting or revealing Russian hostile or covert activities might be a better way to counteract such behavior.

Recommendations

Planners cannot take into account every possible contingency and consideration when thinking about how the Army might be called on to respond to a flashpoint scenario with Russia. However, a few key drivers of escalation in Russia-related contingencies should certainly inform and motivate planning. Territorial contiguity with Russia, former Soviet republic status, and unresolved border issues are characteristics of states likely to be engaged in disputes with Russia. Relatedly, the Army might expect recent disputes to serve as an augur for future ones—and possibly as an augur for flashpoints. Finally, the broader geopolitical context at the time of any interaction with Russia should be considered when gauging the risk of conflict. Whatever its actual weight in the international system, Russia acts like a great power: Geopolitical considerations can drive its decision-making about war and peace.

Another central recommendation to emerge from our research deals with planning for the expected surprise. In the report, we point to several states where flashpoints are possible, not all of which are likely at the top of the Army's planning agenda. The degree to which Army planning is able to take these flashpoint scenarios into account should increase its capacity to forge a timely and effective response. At a minimum, the Army should consider how flashpoint scenarios in the states identified here pose different challenges than the more-mainstream Army crisis contingencies related to Russia.

Looking forward, the Army will want to consider how it might be asked to respond to various potential future flashpoints. Again, future scenarios call for their own analyses and associated wargaming efforts. This approach is the best way to ensure that conflict contingencies are met with as coordinated a response as possible. Factors that the Army will want to take into account are force planning (including manpower and readiness), posture requirements for both short and more-sustained engagements, partner relationships, security assistance arrangements, sustainment efforts, and the challenges associated with the movement of troops.

Even before 2022, Russia did not shy away from militarized disputes, particularly on and near its borders, and occasionally engaged in interactions that escalated to the level of a flashpoint. It also resorts to unconventional methods of influence, such as the use of PMCs or engaging in cyber activities. These behaviors need not precipitate a conflict to which the Army is asked to respond, but they certainly could—as events in Ukraine since 2022 have aptly demonstrated. The more the Army readies itself to deal with potential flashpoints, particularly in areas outside its alliance responsibilities, the easier the task will be should it ever present itself.

Contents

Additional Drivers of Escalation. 83
 Additional Russian Activities That May Drive Escalation 84
 Russian Operations in the Information Environment 86
 Activities of Russian Private Military Contractors. 96
 Conclusion 107

CHAPTER SIX
Conclusions and Implications for the Army 109
 Findings 110
 Implications 113
 Recommendations 114

APPENDIXES
A. Russia's Militarized Disputes, 1992–2010 117
B. Militarized Conflict Case Study: Ukraine Intervention, 2014 123
C. Militarized Conflict Case Study: Russia-Georgia War, 2008 143
D. Militarized Conflict Case Study: Tajik Civil War, 1992–1997 157
E. Militarized Conflict Case Study: Syrian Civil War, 2015–ongoing 171
F. Narratives for Ukrainian and Georgian Disputes 181
G. Coding Justification for PMC Cases 191

Abbreviations. 197
References 199

Figures

Figures

Tables

CHAPTER ONE

Introduction

The research reported here was completed in May 2021.

The years since 2014 have seen the reemergence of Russia as a strategic priority for the United States and the U.S. Army in particular. Even before its 2022 full-scale invasion of Ukraine, Russia had a large number of ongoing and potential disputes with other countries, motivated by a variety of territorial, political, and economic issues. Furthermore, as Moscow has sought to expand its international role, it has increased Russian involvement in civil conflicts, using both overt and covert means. Russian activity in Syria and Libya has raised the prospect that the United States might find itself militarily entangled with Russia in various global hotspots. Our objective in this report is to identify possible Russian conflict scenarios in and near the U.S. Army Europe (USAREUR) area of responsibility (AOR) that could entangle the United States and present distinct military challenges to the Army.

The Army might have an important role in strengthening deterrence or promoting de-escalation near potential points of conflict or preparing to respond to conflicts that place U.S. interests at stake. Before 2022, the policy and research communities focused substantially on contingencies involving the Baltics and other North Atlantic Treaty Organization (NATO) allies, but less attention was paid to the potential conflict risks that might arise from Russian activities elsewhere in and near the USAREUR AOR.

This gap in the Army's appreciation for how it might prepare for and respond to a crisis contingency outside its alliance obligations to NATO members is problematic in light of observed patterns in Russian behavior. Moscow, although increasingly risk-acceptant, has yet to take aggressive military action against a member of the Alliance, which would risk direct retaliation from the United States. Even before 2022, Russia had, however, become increasingly assertive with nonallied states in Europe and nearby regions. These disputes, although not directly invoking U.S. obligations under Article 5 of the Washington Treaty, nonetheless could lead to unintended escalation to conflict with the United States. Differing conflicts in various contexts could place manifold demands on U.S. forces, depending on the nature of the parties involved, the interests at stake, and the timing of such conflicts. With this report, we intend to focus and encourage Army thinking about possible conflicts or flashpoints with Russia in areas and countries beyond U.S. treaty allies.

Research Approach

The research question we address in this report is laden with uncertainty: Where might future Russian flashpoints emerge that could entangle the U.S. Army? This complex question cannot be answered purely probabilistically. Therefore, our research approach involved several lines of effort and differing methodological tools. In short, we synthesized both qualitative and quantitative empirical analyses to gain a better appreciation for where and how a flashpoint could materialize. Our goal was to identify critical factors that drove Russian flashpoints in the past as a way of gaining some analytical insight into where they might occur in the future.[1]

We conducted a qualitative, case study–based analysis of several Russian conflicts to distill the core drivers of these clashes. We complemented this work with a quantitative, statistical evaluation of the correlates of Russian disputes. To the best of our knowledge, this is the first multimethod approach that couples a quantitative analysis (using statistical tools) of Russian militarized disputes and conflicts with a qualitative analysis of these same phenomena (through case studies). As a further qualitative measure, we also analyzed potential additional drivers of flashpoints with Russia, such as cyberattacks, that are not covered specifically by the previous two examinations.

---

[1] For more on the challenges and solutions to crafting research designed to address questions entailing "radical" uncertainty, see John Kay and Mervyn King, *Radical Uncertainty: Decision-Making Beyond the Numbers*, New York: W.W. Norton & Company, 2020.

Qualitative Analysis

We began with an effort to document, and differentiate among, Russia's militarized interactions with other states between 1992 and 2019. We use the term *flashpoint* as a synonym for a *militarized conflict*, which we define as Russia's use of force toward another state (or vice versa) that results in one or more combat deaths, or the overt and explicit military intervention by Russia into an intrastate conflict in support of the recognized government resulting in one or more battle deaths.[2] Flashpoints are distinguished from *militarized disputes*, which are instances involving an explicit threat, display, or use of force by Russia toward the government, official representatives, official forces, property, or territory of another state (or vice versa) that do not result in combat deaths. We refined existing published accounts to derive our data set, which consists of nine conflicts and 54 disputes involving Russia.

Having determined the universe of such cases, we developed an analytical framework to assess the drivers of escalation as a way of determining the causes of Russian flashpoints. The framework consists of a variety of factors (potential drivers of escalation) derived from the existing scholarly literature on crisis escalation and the onset of international war. We brought this analytical framework to bear on four of the conflicts involving Russia. These case studies allowed for a more nuanced assessment of the drivers and nondrivers of escalation in each case. We also assessed several disputes between Russia and the countries Russia later engaged in a conflict to understand why some disputes did not escalate to conflict but others did.

In addition to this analysis of discrete and largely interstate clashes, we also conducted an analysis of other possible drivers of escalatory dynamics. Specifically, we looked at the potential for the deployment of private military contractors (PMCs) and cyber or electromagnetic attacks to lead to a flashpoint.

---

[2] We define Russian intervention into an intrastate conflict on the side of rebels as an *interstate conflict.*

Quantitative Analysis

Our quantitative analysis made use of data on Russian militarized interstate disputes (MIDs) with two distinct purposes in mind. First, we employed statistical modeling techniques to identify those factors most correlated with dispute onset between Russia and 169 other states. For this exercise, we employed data on interactions between pairs of states from 1992 to 2010. These models allowed us to establish the key correlates associated with Russian dispute onset during this 19-year span. Our second aim with the quantitative analysis was to predict which countries, according to specified characteristics, might be most likely to experience a militarized dispute with Russia.[3] To do so, we used the results from our baseline estimates and country-specific data from 2011 to 2018. This predictive analysis allowed us to rank-order countries by the probabilities associated with their likelihood of having a militarized dispute with Russia.

We are aware that this approach has limitations and are careful not to overinterpret the results it produced. Predicting the outbreak of conflict is a difficult endeavor. The relationships and processes generating Russian disputes between 1992 and 2010 are likely not identical to those governing Russia's subsequent behavior, a phenomenon often referred to as *nonstationarity*. We did use country-specific data to develop our predictions. However, lacking more-timely data (post-2010) on Russian militarized deputes, we were unable to fully resolve this issue through our statistical analysis.

Report Structure

In Chapter Two of this report, we present a typology of Russian disputes and conflicts and introduce a data set representing the case universe we examined. We also discuss characteristics of these cases and bin them into categories. (Appendix A provides a full accounting of the disputes.) In Chapter Three, we describe our framework for assessing the causes of escalation to conflict. We used this framework to conduct four case studies of conflicts

---

[3]   Our effort is not unlike that of William G. Howell and Jon C. Pevehouse, *While Dangers Gather: Congressional Checks on Presidential War Powers*, Princeton, N.J.: Princeton University Press, 2007, which examines U.S. incidents of MIDs.

involving Russia, which are summarized in that chapter (and provided in full in Appendixes B–E). We also assess several disputes between Russia and the countries Russia later engaged in a conflict to clarify why some disputes did not escalate to conflict but others did. (Full narratives of these disputes are provided in Appendix F.) In Chapter Four, we offer a quantitative examination of Russian MID onset and predictive analysis of future disputes. In Chapter Five, we present our analysis of additional potential drivers of escalatory dynamics: PMCs and operations in the information environment (OIE). Several cases or vignettes of both are presented (full background on the PMC cases is provided in Appendix G). In Chapter Six, we summarize our findings and offer implications and recommendations for the Army.

CHAPTER TWO

Russia's Disputes and Conflicts

This chapter is largely empirical in nature. It presents definitions, data, and related information on both conflicts and disputes involving Russia between 1992 and 2019. Because the data are largely drawn from the Militarized Interstate Dispute data set,[1] we present some background information on MIDs and describe the modifications that we made to create our list of conflicts and disputes. We also categorize the various disputes involving Russia and present key takeaways.

Background on the Militarized Interstate Disputes Data Set

The MIDs data set compiled by the Correlates of War Project is the most prominent academic effort to collect information about conflicts in which one or more states threaten, display, or use force against one or more other states. The data set spans the years from 1816 to 2010.[2] The Correlates of War Project defines a MID as

> united historical cases of conflict in which the threat, display or use of military force short of war by one member state is explicitly directed toward the government, official representatives, official forces, property, or territory of another state. Disputes are composed of incidents

---

[1] Glenn Palmer, Vito D'Orazio, Michael R. Kenwick, and Roseanne W. McManus, "Updating the Militarized Interstate Dispute Data: A Response to Gibler, Miller, and Little," *International Studies Quarterly*, Vol. 64, No. 2, June 2020.

[2] The version we refer to is version 4 (see Palmer et al., 2020).

that range in intensity from threats to use force to actual combat short of war.[3]

The MIDs data are our point of departure for collecting historical and empirical information about Russian involvement in militarized conflict and disputes between 1992 and 2019.

The (one or more) incidents making up each MID are defined according to several inclusion criteria:

- The event or action must transpire among or be directed toward one or more recognized states.
- The event must not be allowable under international treaties or invited by a target state.
- The event must reflect explicit, nonroutine, and governmentally sanctioned action.
- The event must be an overt action on the part of official military forces or government representatives of a state.
- A militarized incident involving competing territorial claims must take place within the context of a well-defined geographic area.

There are two additional exclusion criteria applicable to incidents. First, military interactions between two states are not coded as separate militarized incidents if the states involved are at war. Second, actions taken by the official forces of one state against private citizens of another state are generally not coded as militarized incidents.[4]

---

[3] The Correlates of War Project uses a threshold of 1,000 battle-related deaths as the level of hostilities that differentiates war from other types of conflict. Daniel M. Jones, Stuart A. Bremer, and J. David Singer, "Militarized Interstate Disputes, 1816–1992: Rationale, Coding Rules, and Empirical Patterns," *Conflict Management and Peace Science*, Vol. 15, No. 2, September 1996, p. 163.

[4] Jones, Bremer, and Singer, 1996.

Modifications to the Militarized Interstate Disputes Data Set

Although it is unique in its comprehensiveness and focus on interstate clashes, the MIDs data set was primarily built to enable cross-country, large-N comparisons. Furthermore, at the time of our analysis, the data set extended only through 2010 (updated data published after this report was completed extend through 2014). We therefore undertook a major effort to refine the existing MIDs data for the purposes of our analysis of Russian flashpoints.

First, we closely examined the 70-odd MIDs involving Russia and corrected several coding errors in the original database.[5] Given the small-n and country-specific nature of our project, this step was critical, because any errors in the data could affect our results. We then devised analytical categories to distinguish among observations in the MIDs. Specifically, we sought to differentiate between incidents in the MIDs that would not qualify as flashpoints (for example, air-to-air intercepts or disagreements over fishing rights) from those that would. We decided to use combat deaths as our key criterion because it is an observable, objective metric of a severe interstate clash.

Definitionally we needed to account for the fact that two of Russia's most-significant foreign interventions involving combat deaths were not interstate disputes (and therefore would not qualify as MIDs). In the cases of Russia's intervention in Tajikistan's and Syria's respective civil wars, Moscow came to the aid of a sitting government in its fight against rebel forces—that is, nonstate actors. Furthermore, two of Russia's most significant military operations involving combat deaths took place after 2010: the interventions in Ukraine and Syria in 2014 and 2015, respectively. These

---

[5] There were three sets of corrections made. First, three of the five entries that began in 1992 listed the Soviet Union, not Russia, as the state involved in the dispute, even though the Soviet Union had already ceased to exist before the end of 1991. We would therefore have overlooked them had we gone strictly by the coding in the data set. Second, several observations in the MIDs either lacked full date information or had erroneous dates. We corrected these after undertaking additional empirical research on the specific disputes. Third, the MID location data set contained numerous errors regarding the location of specified disputes. These were corrected for Tables 2.1 and A.1.

cases were too significant, both in terms of Russia's use of force and in terms of the consequences for the United States, to leave them outside the scope of our analysis.[6]

We therefore categorized our cases as either militarized disputes or militarized conflicts. We define a *militarized dispute* as a case involving an explicit threat, display, or use of force by Russia toward the government, official representatives, official forces, property, or territory of another state (or vice versa) that does not lead to combat deaths. Disputes comprise incidents (i.e., single actions) that range in intensity from threats of use of force to actual combat but only when there are no recorded battle deaths inflicted on or by Russia. Since extending the MIDs data set for Russia through 2019 would have required more resources than were available to our project team, we considered militarized disputes from 1992 to 2010. We further narrowed down the MIDs data set by focusing only on disputes (1) that met the previously stated criterion of not incurring combat deaths and (2) in which Russia was the primary target or initiator (target and initiator states are specified for each observation in the MIDs). This step allowed us to exclude cases in which Russia was one of many states involved in a particular dispute and played a minor role in the course of events.[7]

We defined a *militarized conflict* as Russia's use of force toward another state (or vice versa) that results in one or more combat deaths, or the overt and explicit military intervention by Russia into an intrastate conflict in support of the recognized government resulting in one or more battle deaths.[8] The latter part of the definition allows our project to account for the Tajikistan and Syria interventions. We focused on overt interventions in this context as a way to differentiate between the acknowledged use of force by the state and convert actions, or those undertaken by irregular forces,

---

[6] Several of these conflicts would also have been considered wars by the MID definition (1,000 or more combat deaths).

[7] For example, the MIDs data set lists Russia as one of the states involved in the dispute that ended in the fall of the Taliban in Afghanistan in 2001 because it reinforced its troops stationed in neighboring Tajikistan. But Moscow was certainly not the primary actor in that case.

[8] We treat Russian intervention into an intrastate conflict on the side of rebels as an interstate conflict.

such as PMCs (which we discuss in Chapter Five). We considered all militarized conflicts that occurred between the years 1992 and 2019 to ensure that we took into account the Ukraine and Syria conflicts (thereby adding them to our set of observations).

Data Set

Conflicts

Using these definitions, our case universe consists of nine militarized conflicts and 54 militarized disputes. The nine conflicts are presented in Table 2.1. Overall, Russia engaged in relatively few conflicts between 1992 and 2019. Three of the nine (Abkhazia [Georgia], the Tajik Civil War, and Transnistria [Moldova]) were Russian interventions in civil disputes that had their origin in the collapse of the Soviet Union. Two were relatively minor interstate conflicts that occurred in the context of Russia's two interventions on behalf of its allies against rebel forces: the skirmishes on the Tajikistan-Afghanistan border in 1993–1994 and the Turkish shootdown of a Russian fighter-bomber over Syria in 2015. The former incident involved firefights between Russian border guards stationed in Tajikistan and Tajik rebels who had crossed into—and received support from—Afghanistan. In the latter case, Turkey downed a Russian airframe that had allegedly violated Turkish airspace while attacking Syrian rebel groups. Turkey was the opposing state in another conflict with Russia, a 2000 incident along the Armenia-Turkey border. That frontier is patrolled on the Armenian side by Russian border guards, one of whom was killed in the incident. The remaining three conflicts listed in Table 2.1 are perhaps the most strategically significant: the 2008 Russia-Georgia War, the 2014 intervention in eastern Ukraine, and Russia's involvement in the Syrian Civil War.

Table 2.2 shows additional correlated data on each of the nine conflicts. Several interesting trends emerge. First, with the exception of the two that occurred in Syria, all conflicts took place on the territory of the former Soviet Union. Given that the Soviet Union occupied one-sixth of the world's landmass, this measure does not equate to proximity; however, it does suggest that the borders of the former Soviet Union effectively reflected the outer boundary of the area in which Moscow felt comfortable using force until

TABLE 2.1

Russia's Militarized Conflicts, 1992–2019

Conflict Name	Location	Start Date	Opposing State	Shares Land Border with Russia	Conflict on or Adjacent to Former Soviet Union	Fatalities	Site of Russian Military Facility	Post-New Look (2010)	% of Opposing State's Imports from Russia	% of Opposing State's Exports to Russia
Conflict in Abkhazia	Georgia	June 1992	Georgia	Y	Y	26–100	Y	N	18.51[a]	28.51[a]
Skirmish on the Tajikistan-Afghanistan border	Tajikistan	March 1993	Afghanistan	N	Y	26–100	Y	N	—[b]	—[b]
Russian border guard killed on Armenia-Turkey border	Armenia	June 2000	Turkey	N	Y	1–25	Y	N	7.17	2.33
Russia-Georgia War	Georgia	August 2008	Georgia	Y	Y	101–250	Y	N	11.03	3.7
Tajik Civil War	Tajikistan	May 1992	N/A	N/A	Y	>999	Y	N	N/A	N/A

Table 2.1—Continued

Conflict Name	Location	Start Date	Opposing State	Shares Land Border with Russia	Conflict on or Adjacent to Former Soviet Union	Fatalities	Site of Russian Military Facility	Post-New Look (2010)	% of Opposing State's Imports from Russia	% of Opposing State's Exports to Russia
Intervention in eastern Ukraine	Ukraine	April 2014	Ukraine	Y	Y	>999	Y	Y	23.31	18.18
Intervention in Syrian Civil War	Syria	September 2011	N/A	N/A	N	>999	Y	Y	N/A	N/A
Transnistria conflict	Moldova	March 1992	Moldova	N	Y	501–999	Y	N	46.91[c]	51.18[c]
Turkish shootdown of Russian jet	Syria	November 2015	Turkey	N	N	1–25	Y	Y	9.85	2.5

SOURCE: Start dates, opposing state, and fatalities from MIDs data set, except for the Tajik Civil War and the intervention in the Syrian Civil War, which are from press accounts. Trade data are from WITS, undated. All other columns are the authors' determinations.

NOTE: N/A = not applicable. N/A in the opposing state column means there was no opposing state because Russia was intervening in an intrastate conflict to support the recognized government.

[a] Data are from 1996, the closest available year.

[b] No trade data are available for Afghanistan until 2008.

[c] Data are from 1994, the closest available year.

TABLE 2.2

Summary Data on Russia's 1992–2019 Militarized Conflicts

Characteristic	Number
Total number of conflicts	9
Located in a state hosting a Russian military installation	9
Located on or adjacent to territory of former Soviet Union	7
The opposing state shares a land border with Russia	3 (of 7)[a]
Began post–New Look (2010)	3
Average of opposing states' imports that come from Russia	19%[b]
Average of opposing states' exports that go to Russia	18%[b]

SOURCE: World Bank, undated-c.

[a] Only seven of the nine conflicts were interstate.

[b] Averages are derived from the six opposing states.

2015. Second, only three of the seven interstate militarized conflicts (excluding Tajikistan and Syria) were with countries that share a land border with Russia (Ukraine and Georgia). This suggests that Moscow sees its broader region, rather than just adjacent states, as the zone where its interests are concentrated. Third, the conflicts all occurred in countries where Russia had a military installation. In the former Soviet region, Moscow maintained control over several bases and units after 1992, including the 14th Army in Moldova, several bases in Georgia, the Black Sea Fleet in Crimea, and the 201st Motorized Rifle Division in Tajikistan. So, the presence of a Russian military installation was initially the rule rather than the exception. Today, either with or without host nation consent, Moscow operates facilities on the territory of all but three of the 11 non-Baltic former Soviet republics. Beyond post-Soviet Eurasia, however, Russia has maintained a very light military footprint. After the closure of facilities in Cuba and Vietnam in the early 2000s, Russia's small naval hub and signals intelligence sites in Syria were its only physical military presence outside post-Soviet Eurasia (until it expanded its footprint in Syria in the run-up to and aftermath of its 2015 intervention). Perhaps it is not surprising, then, that the only militarized conflicts involving Russia occurred there. Still, it is striking that Moscow has not engaged in a militarized conflict on the territory of a state where it did not have a base.

Table 2.2 also demonstrates that the majority (six of nine) of the militarized conflicts occurred in the period prior to the implementation of the "New Look" military reforms in Russia.[9] We use 2010 as the year that marked the significant improvements in the Russian military's capabilities following two decades of underinvestment and resistance to reform and modernization. Thus the Russian military frequently engaged in militarized conflicts despite being in the acute stage of its post-Soviet unraveling.[10] All six of these pre–New Look conflicts, tellingly, entailed Russia responding to circumstances not of its own making; none was an unprovoked military adventure. The three post–New Look conflicts, although also arguably driven by Russian responses to exogenously driven changes to the status quo (e.g., the Maidan revolution in Ukraine or the rebellion against the Assad regime), were far more proactive engagements, reflecting Moscow's improved capabilities and increased confidence or an opportunity to test a new and more advanced military. Finally, on average, Moscow's opponent in the conflicts was economically dependent on Russia. The average of the opposing state's imports that came from Russia was 19 percent; moreover, 18 percent of that state's exports went to Russia.[11] One can infer that, despite enjoying economic or trade leverage, Russia has still turned to the use of force to achieve objectives.

Disputes

We grouped the 54 disputes according to the most-prominent and most-visible characteristics. We refer to these groupings as dispute categories. See Table 2.3 for our breakdown of the disputes into these five distinct categories, plus a catch-all category ("other") for those disputes that do not fall under any of the five main categories identified. We created these categories

---

[9] For background on the New Look reforms, see, inter alia, Mikhail Barabanov, ed., *Russia's New Army*, Moscow: Centre for Analysis of Strategies and Technologies, 2011; and Bettina Renz, *Russia's Military Revival*, Cambridge, United Kingdom: Polity, 2018.

[10] See William Odom, *The Collapse of the Soviet Military*, New Haven, Conn.: Yale University Press, 2000.

[11] These averages were calculated according to the year that a conflict began or the closest year for which data is available. Trade data are from World Bank, "World Integrated Trade Solution (WITS)," webpage, undated-c.

TABLE 2.3

Main Dispute Categories

Dispute Category	Number of Disputes
Signaling or Deterrence	14
Defense of or Response to Attacks on Russian Forces or Assets Abroad	13
Commercial or Fishing	8
Border-Related	7
Protracted Regional Conflicts	5
Other	7
Total	54

and sorted the disputes according to the MID narratives, doing additional research when the narratives did not provide adequate information. Each category is defined in greater detail in the following sections.[12]

Here, we present more details about the disputes in the various categories. We numbered the 54 disputes in chronological order (the numbers used in Tables 2.4–2.8 can be matched to the corresponding number in Table A.1). Each table in the rest of this chapter provides at least two additional characteristics for each dispute: whether the opposing state shares a land border with Russia and whether the dispute took place on or adjacent to the territory of the former Soviet Union; in some cases, we provide another characteristic relevant to the specific category.

Signaling or Deterrence

The largest category of disputes, 14 of the total 54, were those related to Russian actions undertaken for signaling or deterrence purposes. These were cases in which Moscow used coercive means to either convey a message or dissuade potential actions that Russian leaders deemed problematic. Eight of these disputes involved Russian air force jets acting assertively (e.g., reportedly straying into a neighbor's airspace or buzzing a U.S. carrier). Others involved demonstrative shows of force seemingly intended to deter the opposing state or, in one case, a nonstate actor (see Table 2.4).

---

[12] All 54 disputes are listed in Table A.1 in Appendix A, along with additional details, such as the corresponding number in the MIDs database. Figure A.1 in Appendix A also provides a visual representation of the opposing states in these disputes.

TABLE 2.4

Summary of Signaling or Deterrence-Related Disputes, 1992–2010

Dispute No.	Opposing State	Dispute Start Date	Opposing State Is U.S. or U.S. Ally	Opposing State Shares Land Border with Russia	Dispute on or Adjacent to the Territory of the Former Soviet Union
3	Sweden	September 1992	N	N	N
12	Ukraine	March 1996	N	Y	Y
24	NATO states	April 1999	Y	N	N
26	United Kingdom (NATO)	June 1999	Y	N	N
27	Norway	June 1999	Y	Y	Y
32	United States	October 2000	Y	N	N
33	Canada and United States	November 2000	Y	N	Y
34	Japan	February 2001	Y	N	Y
35	Norway	February 2001	Y	Y	Y
37	Georgia; Azerbaijan	May 2002	N	N	Y
40	Denmark	April 2003	Y	N	N
47	Japan	January 2006	Y	N	Y
49	Finland	December 2007	N	N	Y
50	Japan	February 2008	Y	N	N

SOURCE: Opposing state and start date are from the MIDs database. Other columns are the authors' determinations.

Eight of the disputes in this category are alleged Russian incursions into or near its neighbors' airspace and, in one case, territorial waters. In only one instance (dispute 12, with Ukraine), Russia alleged a violation of its airspace (and forced a Ukrainian plane to land). Two of the remaining four disputes (24 and 26) represent instances of Russian signaling or deterrence vis-à-vis NATO in the context of the war in Kosovo: One was a naval deployment to the Adriatic; the other was the infamous "dash to Pristina."[13] The dispute with Denmark, a U.S. NATO ally, was the result of an altercation between a Russian naval vessel and a Danish one in the context of the former holding a military exercise in the Baltic Sea. The remaining dispute (37) with Georgia and Azerbaijan resulted from Russian mobilization of reserves on its respective borders with both countries to deter potential cross-border intrusions by Chechen rebels.

Ten of the 14 disputes in this category involved the United States or its allies. This high prevalence suggests that Moscow most often "flexes its muscles" to signal or deter the United States.

Defense of or Response to Attacks on Russian Forces or Assets Abroad

Disputes in this category involve attacks on or defense of Russian soldiers or assets abroad. In terms of Russian assets, we list both military and commercial (or private) assets. For example, Russia and Ukraine were locked in a dispute in July 1994 over the Black Sea Fleet headquarters in Sevastopol, Crimea. In the following days, the garrison in Sevastopol changed hands several times, as Russian and Ukrainian forces tried to maintain control over it.[14] We have identified in this category 13 of the total 54 disputes. Table 2.5 lists them and some of their characteristics.

Disputes Related to Russian Forces or Military Assets Abroad

Nine of the 13 disputes in this category were initiated as a result of Russia's defense of or response to attacks or aggressive actions against Russian forces or military assets abroad. Most of the disputes took place in the 1990s, soon after the collapse of the Soviet Union. The two disputes (1, 9) with Ukraine involved the unsettled questions regarding the assets associated with the Black Sea Fleet. In the remaining seven disputes, Russian troops or military assets were either held or came under attack, with all seven taking place

---

[13] For details, see Anna Maria Brudenell, "Russia's Role in the Kosovo Conflict of 1999," *RUSI Journal*, Vol. 153, No. 1, 2008; and John Norris, *Collision Course: NATO, Russia, and Kosovo*, Santa Barbara, Calif.: Greenwood Publishing Group, Praeger Publishers, 2005.

[14] Tyler Felgenhauer, *Ukraine, Russia, and the Black Sea Fleet Accords*, Princeton, N.J.: Woodrow Wilson School of Public and International Affairs, WWS Case Study 2/99, February 26, 1999.

TABLE 2.5

Summary of Disputes Related to Russian Forces or Assets Abroad, 1992–2010

Dispute No.	Opposing State	Dispute Start Date	Opposing State Shares Land Border with Russia	Dispute on or Adjacent to the Territory of the Former Soviet Union	Involves Assistance to Allied Government
1	Ukraine	July 1992	Y	Y	N
2	Estonia	July 1992	Y	Y	N
5	China	June 1993	Y	N	N
8	Latvia	January 1994	Y	Y	N
9	Ukraine	April 1994	Y	Y	N
10	Afghanistan	December 1994	N	Y	Y
13	Turkey	July 1996	N	Y	Y
17	Afghanistan	May 1997	N	Y	Y
22	Afghanistan	August 1998	N	Y	Y
23	Azerbaijan	March 1999	Y	Y	N
28	Afghanistan	September 1999	N	Y	Y
30	United States	February 2000	N	N	N
41	Sudan	July 2003	N	N	N

SOURCE: Opposing state and start date are from the MIDs database. Other columns are the authors' determinations.

on the territory of the former Soviet Union: two in Baltic states (Estonia and Latvia; disputes 2 and 8), four in Central Asia (on the Afghanistan-Tajikistan and Afghanistan-Uzbekistan borders; disputes 10, 17, 22, and 28), and one in the South Caucasus (in Armenia; dispute 13).

In Central Asia and the South Caucasus, Russian forces came under fire while supporting Moscow's treaty allies. In dispute 13, for example, Russian border guards came under fire from Turkey on the Armenia-Turkey border. Since 1992, as part of Moscow's security assurances to Yerevan, Russian border guards patrol Armenia's border with Turkey and Iran.[15] The remain-

---

[15] "Russian Border Guards' Presence on Border with Turkey Important for Armenia—Premier," *TASS*, July 26, 2018.

ing four disputes (10, 17, 22, and 28) in this subcategory took place along the border with Afghanistan and were primarily related to the threat that Islamic extremists located in Afghanistan posed to Russia's Central Asian allies, Tajikistan and Uzbekistan. The disputes involved attacks from across the border in Afghanistan on Russian troops stationed on the Afghanistan-Tajikistan and Afghanistan-Uzbekistan borders.

Disputes Related to Russian Commercial Assets Abroad

Four disputes related to foreign governments' seizure of commercial or private Russian assets. In three of these disputes (5, 23, and 30), Russian commercial assets were detained or seized under the suspicion that they were engaging in smuggling or were carrying cargo destined to countries under international sanctions or embargo; in one dispute (41), a Russian civilian helicopter carrying humanitarian supplies was impounded by the Sudanese government under the suspicion that it was carrying out a military mission and it did not have the appropriate permission to fly through Sudan.

Commercial or Fishing-Related Disputes

In this category, we have listed disputes that have a commercial character or are related to alleged violations of fishing regulations by Russia or an opposing state. For instance, Russia and Norway have been involved in several disputes related to the rights to operate fishing vessels in the waters around the Spitsbergen (or Svalbard) Arctic archipelago.[16] In another example, in June 2002, a Russian fishing vessel allegedly left international waters crossing into Argentine territorial waters and was fired on (apparently without appropriate warning). Eight of the total 54 disputes fall into this category. They are listed in Table 2.6.

---

[16] Russia's controversy with Norway over the Spitsbergen (or Svalbard) Arctic archipelago dates back to 1977, when Norway established and enforced a 200-nautical-mile maritime Fisheries Protection Zone around the islands. This zone was based on the 1920 Svalbard Treaty granting Norway "'full and absolute sovereignty' over the archipelago." All three disputes with Norway were the result of alleged Russian violations of the zone, which Moscow does not accept. Kristian Åtland and Torbjørn Pedersen, "The Svalbard Archipelago in Russian Security Policy: Overcoming the Legacy of Fear—or Reproducing It?" *European Security*, Vol. 17, No. 2–3, 2008. See also Ivan Stupachenko, "Russia and Norway Clash over Status of Waters Around Spitsbergen/Svalbard," *SeafoodSource*, February 27, 2020.

TABLE 2.6
Summary of Commercial or Fishing-Related Disputes, 1992–2010

Dispute No.	Opposing State	Dispute Start Date	Opposing State Shares Land Border with Russia	Dispute on or Adjacent to the Territory of the Former Soviet Union
7	Poland	June 1993	Y	Y
16	Poland	February 1997	Y	Y
18	United States	August 1997	N	Y
20	Norway	July 1998	Y	Y
38	Argentina	August 2002	N	N
46	Norway	October 2005	Y	Y
52	Norway	July 2008	Y	Y
54	China	February 2009	Y	Y

SOURCE: Opposing state and start date are from the MIDs database. Other columns are the authors' determinations.

Commercial or purely fishing-related disputes are overall relatively few in number, and they involve a diverse set of countries: Argentina, China, Norway, Poland, and the United States. Although Russia experienced militarized disputes associated with fishing rights with other countries (such as Japan), those disputes, at their core, had contested or unsettled maritime demarcations, which ultimately drove the onset of the respective disputes. The disputes in this category were not driven by contested borders and maritime demarcations but by various issues associated with international fishing rights and corresponding practices.

Border-Related Disputes

We classified disputes as border-related if they meet the following three criteria regarding the border between Russia and the opposing state:

1. The border is not agreed or is contested in some fashion by one or both parties.
2. The border is the locus of the dispute.
3. The border disagreement is the clear driver of the dispute.

We considered both land borders and maritime or riverine demarcations between Russia and the opposing state.

We identified seven disputes that meet these criteria. For example, in 1998–1999, Russia reinforced its border with Latvia after the Latvian Prime Minister declared that Riga would unilaterally demarcate the contested area of the shared frontier.[17] Table 2.7 lists the seven disputes along with the type of border subject to the dispute.

All seven border-related disputes were rooted in long-standing historical disagreements over land, maritime, or riverine demarcations with China, Japan, and Latvia, and they predated the collapse of the Soviet Union (including in the Latvian case). The militarized dispute with China on the Amur river was rooted in centuries-long disagreements between Russia and China regarding their common border, dating back to the 1689 Treaty of Nerchinsk, which resulted in the destruction of Russian settlements in the

TABLE 2.7

Summary of Border-Related Disputes, 1992–2010

Dispute No.	Opposing State	Dispute Start Date	Opposing State Shares Land Border with Russia	Dispute on or Adjacent to the Territory of the Former Soviet Union	Type of Border
7	Japan	November 1993	N	N	Maritime
14	China	October 1996	Y	Y	Riverine
15	Japan	October 1996	N	N	Maritime
21	Latvia	August 1998	Y	Y	Land
31	Japan	April 2000	N	N	Maritime
48	Japan	August 2006	N	N	Maritime
53	Japan	January 2009	N	N	Maritime

SOURCE: Opposing state and start date are from the MIDs database. Other columns are the authors' determinations.

---

[17] For a more detailed discussion of the contested Latvian-Russian border dispute, see Claess Levinsson, "The Long Shadow of History: Post-Soviet Border Disputes—The Case of Estonia, Latvia, and Russia," *Connections*, Vol. 5, No. 2, Fall 2006, p. 101. In December 2007, Russia and Latvia finalized their border agreement (Patrick Lannin, "Russia, Latvia Finally Seal Border Treaty," Reuters, December 18, 2007).

Amur basin.[18] (Beijing and Moscow eventually settled their differences and agreed on a final demarcation in 2004.)

The militarized disputes with Japan over the Southern Kuril Islands (referred to as the Northern Territories in Japan) date back to Japan's defeat in the Second World War when the Soviet Union seized the islands.[19] The Soviet seizure of the islands in the final days of the war prevented the two countries from signing a formal peace treaty. Finally, the land dispute with Latvia over the land border demarcation dates back to the 1920 Riga Peace Treaty.[20]

Disputes Related to Protracted Regional Conflicts

This category covers the disputes associated with the protracted regional conflicts in which Russia has been involved since the collapse of the Soviet Union, specifically the conflicts in Transnistria (Moldova), Abkhazia (Georgia), and South Ossetia (Georgia). One such dispute occurred when Russia's 14th Army, which is based in Transnistria, conducted an exercise there in February 1993, prompting an official protest from the Moldovan government. The four Georgia-related disputes (39, 42, 43, 44) took place in the context of disagreements over Abkhazia and South Ossetia, when Russia deployed troops and equipment into the separatist regions. Five of the total 54 disputes are in this category. They are summarized in Table 2.8.

Key Takeaways from Comparisons Across the Five Main Dispute Categories

Most of the signaling or deterrence disputes in which Russia has been involved between 1992 and 2010 (ten of 14 disputes) were with the United States and its allies. The high prevalence of signaling disputes with the

---

[18] For more details, please see Neville Maxwell, "How the Sino-Russian Boundary Conflict Was Finally Settled: From Nerchinsk 1689 to Vladivostok 2005 via Zhenbao Island 1969," in Iwashita Akihiro, ed., *Eager Eyes Fixed on Eurasia*, Vol. 2, *Russia and Its Eastern Edge*, Sopporo, Japan: Slavic Research Center, Hokkaido University, 21st Century COE Program, Slavic Eurasian Studies, No. 16-2, 2007, p. 49.

[19] "Kuril Islands Dispute Between Russia and Japan," *BBC News*, April 29, 2013.

[20] Levinsson, 2006, p. 101.

TABLE 2.8

Summary of Disputes Related to Protracted Regional Conflicts, 1992–2010

Dispute No.	Opposing State	Dispute Start Date	Opposing State Shares Land Border with Russia	Dispute on or Adjacent to the Territory of the Former Soviet Union
4	Moldova	February 1993	N	Y
39	Georgia	February 2003	Y	Y
42	Georgia	September 2003	Y	Y
43	Georgia	April 2004	Y	Y
44	Georgia	March 2005	Y	Y

SOURCE: Opposing state and start date are from the MIDs database. Other columns are the authors' determinations.

United States and its allies suggests that this might be the preferred way for Russia to engage when dealing with a great power competitor. On the one hand, Russia engages in military signaling; on the other, Moscow has thus far been careful to make sure that things do not spiral out of control and do not escalate to conflict level.

The only disputes involving Russia's alliance obligations occurred when its own forces came under threat. Put differently, there were no disputes when Russia came to the aid of an ally when its own forces were not threatened. Given the absence of such episodes, it is fair to say that Moscow has not been particularly invested in carrying out demonstrations of extended deterrence or ally reassurance that involve Russia taking on its allies' concerns as its own.

The number of militarized disputes involving disagreements over land borders is surprisingly low when we consider Russia's numerous unsettled frontiers. Only one of seven border-related disputes in our data set concerns a land border (dispute 21 with Latvia). This is remarkable because Russia's land borders with its neighbors emerged essentially overnight in 1992 in areas where there had been largely meaningless Soviet administrative boundaries before that time. In only one case (with Latvia) did disagreements over the border result in a militarized dispute.

Generally speaking, relative to the number of Russian territorial disagreements with other countries, borders (maritime, land, and riverine) were not a common focal point of militarized disputes. There are over twice as many states with which Russia had such disagreements (15) as there are border-related disputes in our data set.[21] As we noted earlier in this chapter, we define *border-related disputes* quite narrowly: To qualify, the border must be contested in some fashion by one or both parties; the border must be the locus of the dispute; and the border is the clear driver of the dispute. Therefore it is possible that territorial disagreements played a role in the disputes in other categories.

The majority of Russia's disputes stemmed from Moscow's power projection efforts, in the broad sense of the term. Of the 54 disputes, 32 involve Russian forces stationed abroad, Moscow's engagement in protracted regional conflicts, or episodes of signaling or deterrence. All three types of activity amount to projecting military power beyond national borders in one way or another.

---

[21] The 15 states that have had territorial, maritime, or riverine disagreements with Russia in the post-Soviet period are Azerbaijan, China, Estonia, Georgia, Iceland, Iran, Japan, Kazakhstan, Latvia, Lithuania, Norway, Poland, Turkey, Ukraine, and the United States.

CHAPTER THREE

Dispute Onset and Escalation: A Qualitative Analysis

The previous chapter presented data on Russian disputes and conflict. This chapter applies insights from the literature on conflict and escalation to test various potential drivers of escalation on these data. Specifically, we developed a qualitative analytical framework derived from the political science literature and then applied this framework to our data on Russian conflicts and disputes from Chapter Two. The objective was to distill insights from theoretical and empirical studies of conflict to better understand the drivers of Russian flashpoints.

In the first section, we review the literature on dispute onset, escalation, and the causes of conflict to identify 16 potential drivers, ranging from the relative power balance between Russia and its opponent (for interstate conflicts) to the level of domestic political stability in Russia at the time of a conflict. We used this set of possible explanatory variables as a framework to analyze four of the nine identified militarized conflicts involving Russia. This chapter presents a summary of the findings of these four cases and compares them to pinpoint the key drivers.[1] Finally, we use the same 16-factor framework to analyze a subset of the 54 identified militarized disputes from the data set: those involving the same opposing states and similar core disagreements that featured in our conflict cases. In so doing, we identify several factors that are present in the conflicts but absent in the disputes and thus might explain why the former escalated while the latter did not.

---

[1] The detailed studies are provided in full in Appendixes B–E.

Analytical Framework

To determine the drivers of flashpoints involving Russia, we developed a framework to analyze cases of what we referred to and defined in the previous chapter as militarized conflicts. This framework consists of 16 potential drivers of escalation, such as crisis escalation and the onset of international war, derived from the existing scholarly literature on the topic.[2] The literature, much of it derived from large-N studies of the MIDs data set, focuses on various themes and questions associated with the onset and escalation of interstate disputes, crises, and conflicts. The specific issues addressed include why states initiate disputes,[3] the kinds of disagreements that are more likely to result in MIDs,[4] the main predictors of dispute escalation to war,[5] when states are more likely to back down instead of escalate in an interstate dispute,[6] the main predictors of conflict onset and escalation within conflicts,[7] and the main dispute and conflict resolution mechanisms

[2] James D. Fearon, "Rationalist Explanations for War," *International Organization*, Vol. 49, No. 3, Summer 1995; James D. Morrow, "Capabilities, Uncertainty, and Resolve: A Limited Information Model of Crisis Bargaining," *American Journal of Political Science*, Vol. 33, No. 4, November 1989; J. David Singer, "Accounting for International War: The State of the Discipline," *Journal of Peace Research*, Vol. 18, No. 1, 1981; and Alastair Smith, "Testing Theories of Strategic Choice: The Example of Crisis Escalation," *American Journal of Political Science*, Vol. 43, No. 4, October 1999.

[3] William Reed, "A Unified Statistical Model of Conflict Onset and Escalation," *American Journal of Political Science*, Vol. 44, No. 1, January 2000.

[4] Krista E. Wiegand, "Militarized Territorial Disputes: States' Attempts to Transfer Reputation for Resolve," *Journal of Peace Research*, Vol. 48, No. 1, January 2011.

[5] David Kinsella and Bruce Russett, "Conflict Emergence and Escalation in Interactive International Dyads," *Journal of Politics*, Vol. 64, No. 4, November 2002; Peter J. Partell, "Escalation at the Outset: An Analysis of Targets' Responses in Militarized Interstate Disputes," *International Interactions*, Vol. 23, No. 1, 1997; and Paul D. Senese, "Geographical Proximity and Issue Salience: Their Effects on the Escalation of Militarized Interstate Conflict," *Conflict Management and Peace Science*, Vol. 15, No. 2, September 1996.

[6] James D. Fearon, "Domestic Political Audiences and the Escalation of International Disputes," *American Political Science Review*, Vol. 88, No. 3, September 1994b.

[7] Alex Braithwaite, "The Geographic Spread of Militarized Disputes," *Journal of Peace Research*, Vol. 43, No. 5, 2006; and Alex Braithwaite and Douglas Lemke, "Unpacking Escalation," *Conflict Management and Peace Science*, Vol. 28, No. 2, 2011.

on which states rely.[8] We focused on the stream of literature that explores the main predictors of MID onset and escalation, both of which are directly relevant to our research questions.

Our review of the literature revealed considerable lack of clarity regarding the definition of the dependent variable or the phenomenon under consideration. Across various scholarly writings on the topic, the terms *dispute*, *crisis*, *war*, and *conflict* are used interchangeably despite conceptual differences among them.[9] Some studies group *conflict* and *war* together without differentiating between the two.[10] Other studies make no effort to distinguish or define terms.

As noted in the previous chapter, we draw a distinction between the various interstate interactions captured by the MIDs that do not result in combat deaths, which we call *militarized disputes*, and those that result in one or more combat deaths (or interventions in civil wars against nonstate actors), which we call *militarized conflicts*. For the sake of readability, we use *dispute* and *conflict* as shorthand in the remainder of this chapter. Since most conflicts begin as disputes, the drivers of conflict onset are often the same as the factors that drive the escalation of disputes.

Much of the theory and many of the empirical studies do not make the same distinction. Fortunately, however, there is considerable evidence that the drivers of both disputes and conflicts are largely equivalent. Using large-N, cross-country statistical analysis, John R. Oneal and Bruce Russett

---

[8] William J. Dixon, "Third-Party Techniques for Preventing Conflict Escalation and Promoting Peaceful Settlement," *International Organization*, Vol. 50, No. 4, Autumn 1996; Molly M. Melin and Alexandru Grigorescu, "Connecting the Dots: Dispute Resolution and Escalation in a World of Entangled Territorial Claims," *Journal of Conflict Resolution*, Vol. 58, No. 6, 2014.

[9] Bruce Bueno de Mesquita, James D. Morrow, and Ethan R. Zorick, "Capabilities, Perception, and Escalation," *American Political Science Review*, Vol. 91, No. 1, March 1997; Lisa J. Carlson, "A Theory of Escalation and International Conflict," *Journal of Conflict Resolution*, Vol. 39, No. 3, September 1995; and Dixon, 1996.

[10] Paul F. Diehl, "Arms Races and Escalation: A Closer Look," *Journal of Peace Research*, Vol. 20, No. 3, September 1983; Paul Huth, Christopher Gelpi, and D. Scott Bennett, "The Escalation of Great Power Militarized Disputes: Testing Rational Deterrence Theory and Structural Realism," *American Political Science Review*, Vol. 87, No. 3, September 1993.

have argued that the causes of disputes and conflicts "are quite similar."[11] Alex Braithwaite and Douglas Lemke "uncover persistent results that consistently suggest [dispute] Onset and Escalation [to conflict] are linked parts of a single process."[12] Therefore, we were able to leverage a wide swath of literature on dispute onset, escalation, and the causes of war to derive and inform our analytical framework. Moreover, this framework should be useful in understanding the causes of both disputes and conflicts, as we define them in this report.

Drivers

Excluded Factors

We began our effort by compiling an extensive list of the factors identified in the literature as drivers of dispute onset and escalation to conflict or interstate war. Not all of these potential drivers were part of our framework. We ultimately excluded several of them because they were difficult to operationalize and test empirically in the context of our analysis. These included the structure (or polarity) of the international system,[13] power shifts or changes in the relative distribution of power within a dyad,[14] the

---

[11] In their work, Oneal and Russett differentiate among the onset of militarized disputes, the onset of a fatal militarized dispute (one or more deaths), and the onset of war (fatal disputes with at least 1,000 battle deaths). John R. Oneal and Bruce Russett, "Rule of Three, Let It Be? When More Really Is Better," *Conflict Management and Peace Science*, Vol. 22, No. 4, September 2005, p. 299.

[12] Braithwaite and Lemke, 2011, p. 120. Reed echoes this finding: "The central result of this research is that conflict onset and escalation are related processes" (Reed, 2000, p. 92).

[13] Huth, Gelpi, and Bennett, 1993; Smith, 1999.

[14] Smith, 1999; also see Daniel S. Geller, "Power Differentials and War in Rival Dyads," *International Studies Quarterly*, Vol. 37, No. 2, June 1993, p. 173. Conducting a more nuanced analysis of China's territorial disputes, Fravel found that a decline in Chinese relative power in a specific dispute was more likely to result in use of force than when Chinese relative power increased, and he concluded that "[n]egative shifts in bargaining power create incentives for states to use force in territorial disputes" (M. Taylor Fravel, "Power Shifts and Escalation Explaining China's Use of Force in Territorial Disputes," *International Security*, Vol. 32, No. 3, Winter 2007/08, p. 47).

balance of interest within the dyad,[15] and leaders' risk-taking propensity under uncertainty.[16]

The challenges related to operationalizing and testing these factors stemmed from the lack of consensus in the existing scholarly literature regarding the appropriate way to measure or assess a variable (e.g., power or system polarity); disagreements regarding the polarity of the international system (unipolar, bipolar, or multipolar) at a given time; and the difficulty of identifying a clear cut-off date for changes associated with certain variables, such as when transitions in polarity or changes in the balance of interests occurred. For such variables as relative power, change is often gradual over extended periods, and thus difficult to describe with any accuracy.

In addition to operationalization concerns, we excluded several other possible factors mentioned in the literature after a preliminary analysis of their applicability to the Russian context. Examples include Alliance membership and whether the opposing state was an ally or protégé that was economically dependent on Russia.[17] A cursory look at the parties involved in the disputes and conflicts in our data set showed that Russia had been engaged only with Azerbaijan and Georgia as opposing states in disputes or conflicts during a period when they were formally allies.[18] However, both states' alliances with Russia were a sui generis and fleeting consequence of the regional dynamics unleashed by the Soviet collapse; both withdrew from the alliance (known at the time as the Collective Security Treaty) in 1999 after the treaty itself had been in force for only five years. Besides Azer-

---

[15] James D. Fearon, "Signaling Versus the Balance of Power and Interests: An Empirical Test of a Crisis Bargaining Model," *Journal of Conflict Resolution*, Vol. 38, No. 2, *Arms, Alliances, and Cooperation: Formal Models and Empirical Tests*, June 1994a.

[16] Huth, Gelpi, and Bennett, 1993.

[17] Douglas M. Gibler, Steven V. Miller, and Erin K. Little, "An Analysis of the Militarized Interstate Dispute (MID) Dataset, 1816–2001," *International Studies Quarterly*, Vol. 60, No. 4, December 2016; Paul Huth and Bruce Russett, "Deterrence Failure and Crisis Escalation," *International Studies Quarterly*, Vol. 32, No. 1, March 1988, p. 35; and Reed, 2000.

[18] This variable does lend itself more favorably to quantitative measurement in the context of the dyadic MIDs data set, which contains country-year pairs as the basic unit of analysis. See Chapter Four.

baijan and Georgia, no other Russian ally or protégé is the opposing state in any of the disputes or conflicts in our data set.

We grouped the remaining potential drivers of dispute and conflict onset into four categories for our analysis:

1. geographic and territorial drivers
2. drivers associated with the nature of the relationship between the two parties
3. drivers related to threat perceptions and status concerns
4. domestic drivers pertaining to each of the two parties.

Geographic and Territorial Drivers

The literature consistently notes the importance of territorial disagreements in driving escalation.[19] The existence of such disagreements often produces disputes, and when such disputes occur, they are more likely than other types of disputes to result in conflict.[20] This is especially true when the territory under dispute has strategic or ethnic value, when the opposing states are contiguous rivals, or when they have relative power parity.[21] One study found that the odds of disputes involving territorial disagreements are not only over three times more likely to escalate to war than disputes involving nonterritorial issues, but territory-related matters are also more likely to result in recurrent interstate disputes.[22] A study by Gochman, Leng, and Singer further underscores the increased likelihood of escalation when territorial integrity is at stake.[23]

---

[19] We use the term *territorial disagreement* to refer to "a conflicting claim by two or more states over the ownership and control of a piece of land, including islands but excluding maritime demarcation disputes over exclusive economic zones" (Paul K. Huth and Todd L. Allee, *The Democratic Peace and Territorial Conflict in the Twentieth Century*, Cambridge, England: Cambridge University Press, 2003, p. 298).

[20] Braithwaite and Lemke, 2011; Gibler, Miller, and Little, 2016; and Wiegand, 2011.

[21] Wiegand, 2011.

[22] Paul R. Hensel, "Charting a Course to Conflict: Territorial Issues and Interstate Conflict, 1816–1992," *Conflict Management and Peace Science*, Vol. 15, No. 1, 1996.

[23] Charles Gochman and Russell Leng, "Realpolitik and the Road to War: An Analysis of Attributes and Behavior," *International Studies Quarterly*, Vol. 27, No. 1, March 1983; and

Scholars have additionally identified the salience of the territory subject to a dispute as a driver of dispute escalation. In his study of China's use of force in territorial disputes, Fravel specifically identified salience of the territory or "the military, economic, or symbolic value of the land being contested" as one of the key drivers of escalation.[24] His findings are consistent with those of numerous other studies that "establish the domestic salience of territorial issues."[25]

In addition to territorial disagreements and salience, contiguity or proximity of the opposing states has been found to increase the likelihood of escalation. Contiguity of territory translates into a higher number of casualties once force is used in territorial disputes; furthermore, when conflict has started, territorial proximity increases the potential for escalation to war.[26] Research has demonstrated that disputes between two contiguous states engaged in an enduring rivalry are more likely to escalate to war than those between strategic rivals that do not share a border.[27] Contiguity has also been shown to exacerbate disputes driven by territorial disagreements.[28] Other research demonstrates that when "a dispute is contiguous by land to one of the disputants, the likelihood of escalation is increased (and increased even more if both sides are contiguous to the site)."[29] Furthermore, nearby conflicts are harder for citizens and leaders to ignore than faraway conflicts are.[30] For Russia, territorial contiguity is not the only relevant

---

Russell J. Leng and J. David Singer, "Militarized Interstate Crises: The BCOW Typology and Its Applications," *International Studies Quarterly*, Vol. 32, No. 2, June 1988.

[24] Fravel, 2007/08, p. 52.

[25] Gibler, Miller, and Little, 2016, p. 728. See also, Senese, 1996.

[26] Senese, 1996.

[27] John A. Vasquez, "Distinguishing Rivals That Go to War from Those That Do Not: A Quantitative Comparative Case Study of the Two Paths to War," *International Studies Quarterly*, Vol. 40, No. 4, December 1996. Also see Paul F. Diehl, "Contiguity and Military Escalation in Major Power Rivalries, 1816–1980," *Journal of Politics*, Vol. 47, No. 4, November 1985.

[28] Diehl, 1985, p. 1203.

[29] Diehl, 1985, p. 1203.

[30] Karen A. Rasler and William R. Thompson, "Contested Territory, Strategic Rivalries, and Conflict Escalation," *International Studies Quarterly*, Vol. 50, No. 1, March 2006.

measure of geographic proximity. From the beginning of the post-Soviet period, Moscow already had forces stationed on various newly independent states that were not territorially contiguous to it. Therefore, we broadened the definition to reflect not only the former Soviet republics but also territory adjacent to them (e.g., the Tajikistan-Afghanistan border).

To summarize, we identified the following variables that are relevant to our project:

- presence of a territorial disagreement between Russia and an opposing state
- salience of the disputed territory for Russia
- territorial contiguity of the opposing state with Russia
- location of the conflict or dispute on the territory of the former Soviet Union or adjacent to it.

Drivers Associated with the Nature of the Relationship Between Russia and the Opposing State

Previous MID outcomes have been shown to have an impact on the likelihood of the use of force in a dispute.[31] Capitulation in a previous dispute is usually considered to represent a sign of weakness for the yielding state, and Huth, Gelpi, and Bennett argue that capitulation in a previous dispute increases the probability of escalation in a later dispute by 38 percent when the defender (or target state) previously had to back down when confronted by the challenger. Furthermore, when the challenger backed down in a previous dispute with the defender, the probability that an ongoing dispute will escalate decreases by 27 percent.[32]

Given the focus of our analysis, we therefore considered the previous MID outcome for Russia, and particularly whether Russia was the victor in a previous dispute with the opposing state. This factor allowed us to examine the possibility that escalation across disputes and conflicts might be attributable to capitulation by one side in former disputes.

In addition to the outcome of previous disputes, power preponderance within a dyad has been shown to be a driver of dispute onset and escalation

---

[31] For a taxonomy of MID outcomes, see Jones, Bremer, and Singer, 1996, p. 180.

[32] Huth, Gelpi, and Bennett, 1993, p. 618.

to conflict.[33] Although relative strength or power is likely to affect the two parties' decision to escalate a crisis, and conventional wisdom would suggest that states with inferior military capabilities would be more likely to back down, the effect of power preponderance is still debated in the literature, with some research suggesting that conflict escalation is more likely when power parity (rather than preponderance) exists between two rivals.[34] However, more-recent studies suggest that power preponderance is a statistically significant driver of escalation for conflicts with over 250 fatalities.[35] Therefore, we examined the potential explanatory role of power preponderance in our analysis of the cases.

Using our correlates analysis discussed in Chapter Two, we identified two additional variables regarding the nature of Moscow's relations with opposing states that we did not find in the relevant literature but that seem to be correlated with escalation to conflict for Russia: the presence of Russian military facilities and Russian *compatriots* (co-ethnics, Russian citizens, or Russian speakers) on the territory of the state where the conflict takes place. Although there is scant discussion about the relationship between presence of military installations and dispute onset or conflict escalation in the political science literature, recent studies of the Russian interventions in Syria and the operation in Crimea have noted the significance of the existing Russian military facilities prior to the respective interventions. Moreover, our analysis of the conflicts demonstrated that all took place in states where Russian military facilities were located.

The presence of Russian compatriots on the territory of an opposing state could also prove escalatory. This has been a significant concern since the collapse of the Soviet Union, when many compatriots found themselves citizens of newly independent states that were pursuing national projects

---

[33] For details regarding the effect of power preponderance as a driver for onset and escalation of conflict, please see Braithwaite and Lemke, 2011.

[34] Fearon, 1994b, p. 578; Geller, 1993; Gibler, Miller, and Little, 2016; Partell, 1997; Reed, 2000; and Smith, 1999.

[35] Braithwaite and Lemke, 2011, pp. 119–120. The authors operationalize power preponderance as the share of dyadic capabilities possessed by the stronger member of a given dyad. A score of 0.5 reflects perfect equality; a score of 1 indicates complete preponderance by the stronger state.

often explicitly opposed to Russian identity. As David Laitin, the author of the definitive study of these groups, has noted, the fear was that the Kremlin would abuse its "interest in justifying its sphere of influence in the 'near abroad' by claiming that the rights of Russian-speakers are of concern to Russia."[36] The possibility that Moscow could leverage these minority groups has preoccupied policymakers in their newfound countries of citizenship and in the West. Studies of transnational ethnic kin confirm that they are an important catalyst for the initiation and development of internal conflicts, with "ethnic ties that transcend national boundaries . . . [intensifying] conflict by providing sanctuaries as well as human and material resources to the rebels."[37]

To summarize, using our review of the literature and empirical observations after constructing our data set, we identified the following variables or factors related to Russia's relations with opposing states that could plausibly have driven escalation of Russia's disputes:

- the outcome of Russia's previous dispute with the opposing state
- Russia's relative power within the dyad
- the presence of a Russian military installation in the state where the dispute or conflict unfolded
- the presence of Russian compatriots in the territory where the dispute or conflict took place.

Drivers Related to Threat Perceptions and Status Concerns

The literature suggests that, under circumstances of new or increased external security threats (and often in an attempt to avoid crisis escalation and war), states are more likely to adopt measures associated with military

---

[36] David D. Laitin, "Identity in Formation: The Russian-Speaking Nationality in the Post-Soviet Diaspora," *European Journal of Sociology*, Vol. 36, No. 2, 1995, p. 308; also see Igor A. Zevelev, *Russia and Its New Diasporas*, Washington, D.C.: United States Institute of Peace, February 2001; and Fiona Hill, "Mr. Putin and the Art of the Offensive Defense: Approaches to Foreign Policy (Part Two)," Brookings Institution, March 16, 2014.

[37] Mehmet Gurses, "Transnational Ethnic Kin and Civil War Outcomes," *Political Research Quarterly*, Vol. 68, No. 1, March 2015.

buildups, coercive diplomacy, and demonstrations of resolve.[38] For example, Fravel has found that China was more inclined to escalate disputes with its neighbors regarding territorial disagreements when it perceived an increased external threat, such as increased U.S. military and diplomatic support for Taiwan.[39]

Uncertainty about the future has also been identified as a driver of escalation. In an anarchic international system, uncertainty about other states' intentions and their resolve feed into a state's own uncertainty about the future and about its future role in the international system, often impeding cooperation.[40] Under such circumstances, uncertainty about the future can in itself lead to war, even when "no specific conflict of interest or casus belli exists."[41]

Reputational costs have also been found to play a role in driving escalatory dynamics. Studies have shown that states are likely to initiate militarized disputes in an attempt to send a costly and, hence, credible signal.[42] This signal is often aimed not only at the opposing state in a specific dispute but also at other potential adversaries; escalation is often intended to project a reputation for resolve and thus to deter other adversaries.[43] Other research

---

[38] Sambuddha Ghatak, Aaron Gold, and Brandon C. Prins, "External Threat and the Limits of Democratic Pacifism," *Conflict Management and Peace Science*, Vol. 34, No. 2, March 2017; and John A. Vasquez, *The War Puzzle Revisited*, Cambridge, United Kingdom: Cambridge University Press, 2009, pp. 185, 193–194, 232, and 235.

[39] Fravel, 2007/08, pp. 52, 58.

[40] Charles Glaser, "Political Consequences of Military Strategy: Expanding and Refining the Spiral and Deterrence Models," *World Politics*, Vol. 44, No. 4, July 1992; Charles Glaser, "Realists as Optimists: Cooperation as Self-Help," in Michael E. Brown, Owen R. Coté, Jr., Sean M. Lynn-Jones, and Steven E. Miller, eds., *Theories of War and Peace: An International Security Reader*, Cambridge, Mass.: MIT Press, 1998, p. 100; Robert Jervis, "Cooperation under the Security Dilemma," *World Politics*, Vol. 30, No. 2, January 1978; Morrow, 1989; and Kenneth Waltz, *Theory of International Politics*, New York: Random House, 1979, p. 105.

[41] Fravel, 2007/08, p. 48.

[42] Kinsella and Russett, 2002.

[43] Wiegand, 2011.

has shown that rivals might escalate a territorial dispute and use force "to signal general resolve or coerce the rival over another issue."[44]

Finally, drawing on power transition theory, some scholars have argued that states satisfied with the nature of the international system are less likely to be in conflict than states that are dissatisfied with the system. Kugler and Lemke, for example, make the case that pairs of jointly satisfied states are more likely than dissatisfied states to avoid conflict with each other because there are few or no contentious reasons for a clash.[45] When disagreements between such states do occur, the affinity between them is expected to be so high that they find negotiated settlements.[46] Looking at individual states, Jonathan Renshon argues that "status dissatisfaction" is "significantly associated with an increased probability of war and militarized interstate dispute (MID) initiation."[47]

The variables identified in this category are therefore as follows:

- Russia's perception of new or increased external security threats
- Russia's experiencing increased acute uncertainty about the future
  - We focused on increased acute uncertainty to isolate circumstances in which Russia was experiencing a level of uncertainty that is so much higher than the norm that it could plausibly drive escalatory behavior. This factor allows us to isolate moments when uncertainty about future developments that affect core security concerns could drive decisions about war and peace. Although uncertainty is

---

[44] Fravel, 2007/08, p. 79. See also Rasler and Thompson, 2006.

[45] Jacek Kugler and Douglas Lemke, "The Power Transition Research Program: Assessing Theoretical and Empirical Advances," in Manus I. Midlarsky, ed., *Handbook of War Studies II*, Ann Arbor, Mich.: University of Michigan Press, 2000, p. 513.

[46] Douglas Lemke, *Regions of War and Peace*, New York: Cambridge University Press, 2002; Douglas Lemke and William Reed, "Regime Types and Status Quo Evaluations: Power Transition Theory and the Democratic Peace," *International Interactions*, Vol. 22, No. 2, 1996; and A. F. K. Organski and Jacek Kugler, *The War Ledger*, Chicago: University of Chicago Press, 1980. More-recent research findings show that although joint satisfaction as onset predictor remains unchanged, it is no longer a significant predictor of escalation. See Braithwaite and Lemke, 2011.

[47] Jonathan Renshon, "Status Deficits and War," *International Organization*, Vol. 70, No. 3, 2016, p. 513.

a constant, certain events (for example, a change of government in a neighboring state that could dramatically shift that state's foreign policy) create windows of acute uncertainty that might spark a conflict.

- reputational costs for Russia for nonintervention
  - We operationalized the factor of reputation by seeking to identify potential costs to Moscow for not engaging in the conflict. It is more difficult to pinpoint whether demonstrating resolve was a driver of a conflict. Therefore, we focused on identifying the potential costs of inaction.
- Russia's satisfaction with its place in the international system.
  - We chose to focus this variable on Russia itself, as opposed to dyadically. Russia's opponents in its conflicts tend not to be major powers, and thus are less likely to be driven by concerns about the international system. Moreover, focusing only on Russia allowed for cross-case comparison and better fit the objectives of the project. Determining when Russia was and was not satisfied with its place in the international system is a subjective exercise. One could make the case that Moscow has never been fully satisfied with its place in the international system after the collapse of the Soviet Union. However, for much of this period, one can make the case that Russian resentment was not a primary driver of the country's foreign policy, in part because the prospect of some sort of alignment with the West still seemed plausible to many. Until the mid-2000s, for example, senior Russian officials treated possible Russian membership in the European Union (EU) or even NATO as a plausible scenario. Around that time, this hopeful view of the future was displaced by a drive to push back on what Russia saw as unilateral attempts by the United States to impose its will on the world. We chose 2007 (the year of Russian President Vladimir Putin's infamous "Munich speech," in which he outlined Russia's deep dissatisfaction with its place in the system and the system itself) as the year after which this factor was present for Russia.[48] The speech itself was a marker of what Renshon

---

[48] See President of Russia, "Speech and the Following Discussion at the Munich Conference on Security Policy," transcript, Munich, February 10, 2007.

calls "a heightened concern for status triggered by status deficits within a given status community," that came after a series of events Putin discussed in the speech: the NATO bombing of Kosovo, the U.S. invasion of Iraq, the Color Revolutions in Ukraine and Georgia, the U.S. withdrawal from the Anti-Ballistic Missile Treaty, etc.[49] Although the dissatisfaction had thus been building up for years, the speech serves as a clear marker of a point of no return to the previous era of Russian aspirations to essentially join the U.S.-led order.

Drivers Related to Domestic Political Dynamics

A state that experiences domestic instability or other domestic threats is likely to fear that its opponent in a dispute could take advantage of its internal situation. Domestic turbulence can exacerbate a state's perception of vulnerability vis-à-vis the opposing state in the dispute.[50] Fearing the loss of the first mover's advantage, the state experiencing domestic instability is likely to preemptively escalate the dispute.[51] For example, during the economic crisis and domestic instability of the late 1950s and early 1960s, China escalated its border disagreements with India and the Soviet Union partly out of concern about their possible moves to take advantage of the internal weakness that China was experiencing.[52] Along similar lines, a state might escalate a dispute—usually involving an issue with domestic political resonance—to rally society around the flag; divert attention from other, more-pressing internal issues; or marginalize political opposition.[53]

---

[49] Renshon, 2016, p. 514.

[50] Jack S. Levy, "The Diversionary Theory of War: A Critique," in Manus I. Midlarsky, ed., *Handbook of War Studies*, Boston: Unwin Hyman, 1989; and John E. Mueller, *War, Presidents, and Public Opinion*, New York: John Wiley & Sons, 1973. Also see the discussion in Fravel, 2007/08, p. 52.

[51] For details see Zeev Maoz, "Resolve, Capabilities, and the Outcomes of Interstate Disputes, 1816–1976," *Journal of Conflict Resolution*, Vol. 27, No. 2, June 1983.

[52] Fravel, 2007/08, pp. 69–70, 76–77.

[53] Brett Ashley Leeds and David R. Davis, "Domestic Political Vulnerability and International Disputes," *Journal of Conflict Resolution*, Vol. 41, No. 6, 1997; and Clifton T. Morgan and Kenneth N. Bickers, "Domestic Discontent and the External Use of Force," *Journal of Conflict Resolution*, Vol. 36, No. 1, March 1992.

Although the link between domestic instability and escalation is relatively well established, the relationship between domestic regime type and escalation is more ambiguous in the literature. Studies have shown that when two opposing states are democracies, dispute onset is less likely. According to James Fearon, democratic states have stronger domestic audience costs that make them less likely to back down and thus render their public signaling of intentions more credible. This credibility of commitment leads to an amelioration of the security dilemma between democratic states. By contrast, Fearon argued, authoritarian states, which have lower domestic audience costs, are less likely to credibly bind themselves to a specific course of action.[54] However, once two democracies engage in a dispute, escalation to conflict becomes more likely, although this finding has been questioned by other studies.[55]

Since Russia has never been considered a consolidated democracy in the post-Soviet period, we sought insights from the literature on the role of the factor that could vary: regime type of the opposing state. Gleditsch and Hegre show that mixed dyads of a democratic and nondemocratic state are more likely to experience war than either democratic or nondemocratic dyads.[56] A dispute between Russia and a state with a democratic regime would have resulted in a mixed dyad, which could have been a driver of escalation.

To summarize, we identified the following possible domestic drivers of escalation:

- the level of political stability in Russia
- the level of political stability in the opposing state
- the regime type of the opposing state. (Since Russia was more or less consistently non-democratic, the regime type of the opposing state would determine whether the resulting dyad was a mixed one and more prone

---

[54] Fearon, 1994b. Also see Kinsella and Russett, 2002. Recent work by Jessica Chen Weiss has demonstrated that authoritarian regimes also face domestic audience costs in their foreign policies (Jessica Chen Weiss, *Powerful Patriots: Nationalist Protest in China's Foreign Relations*, New York: Oxford University Press, 2014).

[55] Gibler, Miller, and Little, 2016; and Palmer et al., 2020.

[56] Nils Petter Gleditsch and Håvard Hegre, "Peace and Democracy: Three Levels of Analysis," *Journal of Conflict Resolution*, Vol. 41, No. 2, April 1997, p. 283.

to escalation or a nondemocratic one and less inclined toward escalation. We used the Polity IV democracy scores for this purpose.[57])

Summary: Potential Drivers of Escalation

Table 3.1 summarizes the 16 factors that we applied and tested on four cases of conflict and six disputes.

TABLE 3.1

Framework: Potential Drivers of Dispute Onset and Escalation

Factor Number	Driver
1	Is this a territorial issue for Russia?
2	Is this a territorial issue for the opposing state?
3	Is the disputed territory salient to Russia?
4	Is the dispute or conflict located in a state hosting a Russian military facility?
5	Is the dispute or conflict located on or adjacent to the territory of the former Soviet Union?
6	Does the opposing state share a land border with Russia?
7	Is Russia experiencing acute increased uncertainty about the future?
8	Does Russia face or perceive new or increased external security threats?
9	Is Russia experiencing domestic political instability?
10	Is the opposing state experiencing domestic political instability?
11	Are there reputational costs for Russia for nonintervention?
12	Was Russia the victor in a previous dispute with the opposing state?
13	Is power preponderance within the dyad in Russia's favor?
14	Is Russia dissatisfied with its place in the international system?
15	Is the opposing state a democracy?
16	Are Russian compatriots present on the territory of the state where the dispute or conflict takes place?

[57] Polity IV Project, "Polity IV Individual Country Regime Trends, 1946–2013," webpage, June 6, 2014.

As noted earlier in this chapter, our definition of militarized conflicts encompasses interventions in civil conflicts on behalf of a government against a rebellion. By definition, these events are not interstate, dyadic contests because the opponent of the intervening state is a nonstate actor. Therefore several of these potential drivers of escalation—those that refer to the nature of a bilateral relationship or the characteristics of the opposing state (factors 1, 2, 6, 10, 12, 13, and 15 in Table 3.1)—are not applicable to such conflicts.[58]

Analysis of Russia's Militarized Conflicts, 1992–2019

We chose four of the nine identified conflicts involving Russia for examination using this framework based on several criteria. First, to focus on the most-significant cases of use of force, we excluded the four conflicts with under 100 fatalities: Russia's 1993 intervention in Abkhazia;[59] the two clashes with Turkey that each resulted in a single combat death; and the 1993 cross-border incident with Afghanistan. These cases were generally short-lived and often involved tactical situations created by states other than Russia. Thus, the ability of these cases to assist in our understanding of future Russian flashpoints is somewhat limited. Second, of all the conflicts, the Russia-Georgia War, the 2014 intervention in Ukraine, and the intervention in the Syrian Civil War stood out as the most strategically significant; they each transformed perceptions of Russia as an international actor. They were also the most militarily taxing of the cases in terms of resources (and capabilities, at least in the case of Syria). Third, of the two remaining cases, the Tajik Civil War stood out as the only case of a Russian intervention in post-Soviet Eurasia to prop up a sitting friendly regime against a rebellion, whereas the Transnistrian War was akin to the Ukraine

---

[58] Although there is a rich literature on drivers of interventions in civil conflicts, it does not draw the distinction between interventions on behalf of the sitting government or a rebel movement. Since we treat interventions in favor of rebellions as interstate conflicts in this study, insights from that literature would have been difficult to operationalize across our cases.

[59] The civil conflict between Georgians and Abkhazians certainly led to more than 100 fatalities; Russia's intervention did not.

and Georgia conflicts in that it involved Moscow acting to support a rebel force against a central government. Given current concerns about a similar Russian intervention in Belarus, where Russia-friendly regimes face significant domestic opposition, we chose to examine the Tajik case.

Here, we present the key conclusions regarding drivers of conflicts from each of the four case studies. (The detailed case studies we conducted on the four conflicts are provided in full in Appendixes B–E.) We then compare the cases to understand better the drivers of conflict involving Russia.

Case Study Summaries
Intervention in Ukraine, 2014

This case study (see Appendix B) covers the Russian intervention in the eastern Ukrainian macroregion of the Donbas, beginning with the first recorded combat deaths in April 2014. Our analysis found 11 of the 16 possible drivers of escalation to have played roles of varying importance. We divided the factors that were important drivers of escalation in this case into structural factors and proximate triggers. Ukraine's geographic proximity to Russia, its status as the largest and most strategically important former Soviet republic, the enduring salience of the Donbas for Russia (given the multitude of cross-border ties), Russia's military strength relative to Ukraine, and the existence of Russian compatriots in the Donbas have all been immutable facts of life since 1992. Russian dissatisfaction with its position in the international system, although a more recent phenomenon, also predates the conflict. These longer-term structural factors established the preconditions for conflict between Ukraine and Russia. That said, they were not sufficient. What changed in 2014 were the intensification of Russian threat perceptions, a dramatic intensification of Russian uncertainty about the future sparked by the Maidan Revolution, Ukrainian domestic political instability, the prospect of threats to Ukraine's territorial integrity, and significant reputational costs for Moscow for nonintervention.

Russia-Georgia War, 2008

The August 2008 war (see Appendix C) was a complex and multidimensional event in which seven of the 16 factors played a role in the outcome. The conflict was clearly a territorial issue for Georgia. Since its independence, the country had never established control over all of its internationally recognized territory, and President Mikheil Saakashvili, who came to power in 2003, sought to restore Tbilisi's authority. The long-standing presence of Russian peacekeepers in South Ossetia allowed Russia to carry out military buildups in the region, and their presence also intensified escalation when those peacekeepers came in harm's way as Georgian troops advanced into the breakaway region. The conflict was also structurally driven by the proximity of the two opposing states and Russia's determination to play a role of regional hegemon. A series of decisions by the United States and its European allies in early 2008 created a significant degree of Russian uncertainty about the future, specifically regarding the prospect of Georgia's NATO membership. That prospect also sharpened threat perceptions in Moscow, which continues to see NATO presence near its borders as a grave challenge to its security. After the Georgian operation began, the reputational costs to Moscow for not responding with overwhelming force became a significant driver. Had Tbilisi succeeded, Russia's reputation as a regional hegemon and patron of separatist regions would have been called into question. Because its peacekeepers were under attack, there also would have been reputational costs to cover at home. All of this was occurring in the post-"Munich speech" context of Russian discontent with its place in the international system and the resulting foreign policy assertiveness. Finally, the vast majority of South Ossetians had Russian passports; thus, Russia's self-declared responsibility to protect them proved another escalatory dynamic.

Tajik Civil War, 1992–1997

As the Soviet Union collapsed, a bloody civil war broke out in newly independent Tajikistan, pitting former Communist Party functionaries against a rebel movement that included extremist elements (see Appendix D). Russia, which took control of several Soviet military facilities in Tajikistan, found itself gradually embroiled in the conflict, slowly increasing its involvement over 15 months between May 1992 and July 1993. Russia's military intervention in the Tajik Civil War had seven main drivers (of a possible total of nine, because the rest do not apply to interventions on behalf of the sitting government). First, Tajikistan's location in geographic proximity to the Russian Federation and its status as a former Soviet republic contributed to Moscow's view that the civil war had implications for its own security, even though the two states do not share a land border. Second, the presence of Russian military units and border guards on the territory of Tajikistan when the civil war started entangled Moscow in the conflict from the start, particularly given the rebellion's cross-border activities and attacks on Russian forces. Third, the external threat of transnational Islamic extremism spreading from Tajikistan to Russia and other Central Asian governments was a significant driver of Russia's involvement. Fourth, the presence of a sizable ethnic Russian community in Tajikistan also drove Moscow to greater entanglement in the civil war once some members of the community were caught in the crossfire. Fifth, the recent dissolution of the Soviet Union created significant uncertainty about the future of the regional order. Some believed that if the weak, newly independent states, such as Tajikistan, were to fall to rebellions, the rest of the region might follow suit. Sixth, there were major potential reputational costs for refusing to intervene for Moscow, calling into question its status as a regional hegemon. Finally, the domestic political instability in Russia in 1992–1994 helped explain the initial ad hoc escalation of Moscow's involvement, as officers on the ground took matters into their own hands without clear guidance from home.

Syrian Civil War, 2015–ongoing

Russia's intervention in September 2015 in Syria's civil war surprised many observers. Most had assumed that Moscow had neither the interest nor the capabilities to conduct a combat operation beyond post-Soviet Eurasia. Although it is true that Russia had been assisting the regime of Bashar al-Assad since the outbreak of hostilities in Syria in March 2011, the air operation marked a distinct escalation of its involvement (see Appendix E). There were four central drivers of this escalation of Russia's involvement in the Syrian Civil War. First, the situation on the ground in Syria fostered acute increased uncertainty among Russian decisionmakers about the future. More specifically, a key cause of the escalation was the potential that, barring an external military intervention, Assad would fall within months, perhaps even weeks, and that he would likely be replaced by a regime hostile to Russian interests—or by no regime at all, a situation that Moscow saw as potentially empowering extremist elements. Assad's potential ouster was associated with two additional drivers: external threats and reputational costs for nonintervention. Moscow appeared to believe that Assad's fall would legitimize what it sees as a U.S. policy of ousting governments that do not comply with Washington's wishes, a policy that Moscow fears could be applied to Russia itself in the future. To some extent, then, Russia's leadership might have seen escalation in Syria as a type of forward defense of the homeland. Additionally, the Kremlin saw links between extremists in Russia and their counterparts in Syria that created a terrorist threat. Moreover, had Moscow not intervened, it risked losing its only client in the Middle East, along with its sole remaining military facilities, and thus its regional standing. The fourth driver of the escalation of Russia's involvement in Syria was Russia's dissatisfaction with its place in the international system—which was palpably more acute even than the post-2007 "Munich speech" norm. Following the annexation of Crimea and invasion of eastern Ukraine in 2014, the West had attempted to isolate Russia diplomatically. Moscow appeared to believe that intervening in the Syrian Civil War would compel the West to deal with Russia and would represent a breakdown of the West's attempted diplomatic blockade.

Cross-Case Comparison

Table 3.2 presents the first step of the analysis in all of the case studies, testing whether the given factor from our framework is present in the given conflict, absent, or not applicable (in cases of non-interstate conflicts). It does not show the results of the analysis; instead, it narrows the list of factors that could be relevant to the respective cases by assessing whether a given potential driver was present. For example, the 2014 Ukraine conflict did not entail a territorial issue for Russia as we define it (i.e., Russia's own territorial integrity was not under threat). We therefore coded that box as a no. However, that conflict certainly was a territorial issue for Ukraine, the opposing state in this case. Many of the potential drivers are not applicable to Tajikistan and Syria because those cases are not interstate conflicts.

Table 3.3 demonstrates the results of the case study analysis. The factors marked "Y" (for yes) are those deemed to have been drivers of escalation in the given case. "N" markings (for no) are indications that a factor was present in a given case but was not a driver of escalation. For example, Ukraine and Georgia were both democratic at the time of the conflicts, but

TABLE 3.2

Presence of Potential Drivers in the Cases

Factor No.	Driver	Ukraine Conflict (2014)	Georgia Conflict	Tajikistan Conflict	Syria Conflict
1	Is this a territorial issue for Russia?	N	N	N/A	N/A
2	Is this a territorial issue for the opposing state?	Y	Y	N/A	N/A
3	Is the disputed territory salient to Russia?	Y	Y	N/A	N/A
4	Is the conflict located in a state hosting a Russian military installation?	Y[a]	Y	Y	Y
5	Is the conflict located on or adjacent to territory of former Soviet Union?	Y	Y	Y	N
6	Does the opposing state share a land border with Russia?	Y	Y	N	N
7	Is Russia experiencing acute increased uncertainty about the future?	Y	Y	Y	Y
8	Does Russia face or perceive new or increased external security threats?	Y	Y	Y	Y
9	Does Russia face domestic political instability?	N	N	Y	N
10	Is the opposing state experiencing internal instability?	Y	N	N/A	N/A
11	Were there reputational costs for Russia for nonintervention?	Y	Y	Y	Y
12	Was Russia the victor in a previous dispute with the opposing state?[b]	N	N	N/A	N/A
13	Is power preponderance within the dyad in Russia's favor?	Y	Y	N/A	N/A
14	Is Russia dissatisfied with its place in the international system?	Y	Y	N	Y
15	Is the opposing state a democracy?[c]	Y	Y	N/A	N/A
16	Are Russian compatriots present on the territory where the conflict takes place?	Y	Y	Y	N

NOTE: Y = factor is present; N = factor is absent; N/A = factor not applicable.

[a] Russia had a military facility in Crimea, internationally recognized as Ukrainian territory at the time of the conflict.

[b] This variable was determined by the coding of the outcomes of previous MIDs between Russia and the respective opposing state in the MIDs data set.

[c] We used Polity IV scores to determine the nature of the opposing state's regime in the year the conflict began (Polity IV Project, 2014).

TABLE 3.3

Assessment of Potential Drivers of Escalation

Factor No.	Driver	Ukraine Conflict (2014)	Georgia Conflict	Tajikistan Conflict	Syria Conflict
1	Territorial issue for Russia	A	A	N/A	N/A
2	Territorial issue for the opposing state	Y	Y	N/A	N/A
3	Disputed territory is salient to Russia	Y	N	N/A	N/A
4	Russian military installation present on the territory of the state where conflict unfolds[a]	N	Y	Y	N
5	Dispute or conflict located on or adjacent to former Soviet republic	Y	Y	Y	A
6	The opposing state shares a land border with Russia	Y	Y	A	A
7	Russia experiences acute increased uncertainty about the future	Y	Y	Y	Y
8	Russia faces or perceives new or increased external security threats	Y	Y	Y	Y
9	Russia faces domestic political instability	A	A	Y	A
10	The opposing state experiences internal instability	Y	A	N/A	N/A
11	Reputational costs for Russia for nonintervention	Y	Y	Y	Y
12	Russia was the victor in a previous dispute with the opposing state	A	A	N/A	N/A
13	Power preponderance within the dyad was in Russia's favor	Y	N	N/A	N/A
14	Russia was dissatisfied with its place in the international system	Y	Y	A	Y
15	The opposing state is a democracy	N	N	N/A	N/A
16	Presence of Russian compatriots	Y	Y	Y	A

NOTE: Y = factor is a driver; N = factor is present, but not a driver; A = factor is absent; N/A = factor not applicable.

[a] Russia had a military facility in Crimea, internationally recognized as Ukrainian territory at the time of the conflict.

the nature of their regime did not play a role in driving escalation. "A" (for absent) coding replicates the negative coding from Table 2.2.

Drawing on this comparison, three categories of drivers of flashpoints seem to have been central across these diverse cases. First, in all the conflicts that took place near Russia (all but Syria), proximity itself played a role in driving escalation. This dynamic underscores not only the reality that Moscow has been capable and willing to use force in its immediate environs, but also the centrality of the proximity in driving Russian flashpoints. Second, broader geopolitical factors—Russia's dissatisfaction with its place in the international system, acute uncertainty about the future, and reputational costs—were central to the escalatory dynamic in all four conflicts. This was true for both interstate clashes and interventions in civil conflicts. Finally, external threats to Russia were often the immediate trigger. External security threats were identified as drivers of escalation in all of the conflicts. Taken together, these observations reinforce the view of Russia as a status-seeking, geopolitically minded, but predominately regional power— or at least a power that sees its immediate environs as the primary source of security threats. Furthermore, the centrality of external threats to Russia's calculus across the cases suggests that Russia is driven to flashpoints by the perception of imminent potential losses, not by an expansionist instinct.[60]

Two additional, though more tenuous, findings emerge from the comparison. Although we only examined two cases of interstate conflict, both opposing states were considered democratic at the time, but in neither case could we identify a causal role that regime type played in driving escalation. This finding contrasts with both the theory regarding authoritarian-democratic dyads cited earlier in this chapter and the commonly encountered claim that Russia goes to war in order to prevent "democratic contagion."[61] Second, we had assumed that the presence of a Russian military facility in the state where a conflict was located could drive escalation by facilitating Moscow's intervention. However, in the particular circum-

---

[60] See Brian Lampert, "Putin's Prospects: Vladimir Putin's Decision-Making Through the Lens of Prospect Theory," *Small Wars Journal*, February 15, 2016.

[61] See, for example, Thomas Ambrosio, "Insulating Russia from a Color Revolution: How the Kremlin Resists Regional Democratic Trends," *Democratization*, Vol. 14, No. 2, 2007.

stances of the relevant cases (Georgia and Tajikistan), these facilities deepened Russia's involvement because they put its forces in harm's way, thus pushing Moscow to escalate, rather than serving as an asset that allowed for enhanced power projection.

Analysis of Russia's Militarized Disputes

For *militarized disputes* (interstate discord that does not rise to the level where combat deaths are recorded), the nature of the event often tends to be closely related to the primary driver of dispute onset. Therefore, the five categories of disputes discussed in Chapter Two are, in many cases, synonymous with the key factor in dispute onset. (The proximate cause of the border-related disputes, for example, was a territorial disagreement between Russia and the opposing state.)

However, we did not apply the 16-factor framework to all 54 disputes, for two primary reasons. First, many of the 54 disputes, such as those involving fishing issues, are rather unidimensional events lacking the complexity and multicausality often involved in the onset of war; therefore, many disputes are not appropriate for comprehensive qualitative case studies. Second, these disputes are, in essence, a null set—none of the 54 escalated to conflict, in light of our project's focus on flashpoints (i.e., militarized conflicts). Therefore, the analytical leverage on our dependent variable would be limited.

However, there is a methodological benefit to avoiding selection on the dependent variable: Our causal inferences about the drivers of Russian flashpoints will be more robust if we examine some disputes that did not escalate to conflict. But comparing the Russia-Georgia War with, for example, a fishing dispute with Argentina would make little sense practically or methodologically. Therefore, we chose to examine disputes between Russia and the two states with which it later had conflicts—Ukraine and Georgia.[62] Among the possible disputes with those two countries, we focused on those

---

[62] In addition to Georgia and Ukraine, Russia had both disputes and conflicts with Turkey. However, we chose not to focus on Turkey because the conflicts involved only one combat death and were much less strategically significant than the Georgia and Ukraine conflicts.

that involved issues broadly similar to the conflicts (e.g., for Georgia, disputes prior to the 2008 war that centered on Russian involvement in the separatist regions). Holding constant the opposing state and the issue at stake allowed us to better understand why the latter episodes escalated to conflict.

Comparing Disputes with Ukraine and Georgia That Did Not Escalate to Conflict with Those That Did

We applied the 16-factor framework to test for the presence of potential drivers of escalation for six disputes: two Ukraine-related disputes (1 and 9) and four Georgia-related disputes (39, 42, 43, and 44).[63] These disputes involved broadly similar issues to the subsequent conflicts that Russia had with both states. (Descriptive narratives of these disputes are provided in Appendix F.) We compare these disputes, which did not escalate to conflict, with the two disputes that did escalate to conflict—the 2014 invasion of Ukraine and the 2008 Russia-Georgia War—by isolating factors that are absent in the disputes but present in the conflicts (i.e., those that correlate with escalation).

Ukraine Cases

Disputes 1 and 9 both relate to the disagreement between Russia and Ukraine regarding the division and control of the Black Sea Fleet located in Crimea, Ukraine, after the collapse of the Soviet Union. Both countries claimed unified and total control over the fleet. Ukraine claimed control over the Black Sea Fleet on the basis of the fleet's location on what at the time was undisputed Ukrainian territory (Crimea); Russia made the same claims based on the fleet's nuclear forces, which—Russia argued—should

---

[63] We excluded the three other disputes involving Ukraine (disputes 12, 45, and 51) from this analysis. The first dispute (12) was unrelated to Crimea or the Donbas. For the other two (45 and 51), we lack adequate information about the drivers. We also excluded the three other militarized disputes with Georgia (disputes 19, 29, and 36) because they were unrelated to the Russia-backed separatist regions. Dispute 19 was recorded in 1997 when Russia moved a border post into Georgian territory to curtail alcohol smuggling across the border. Disputes 29 and 36 are both associated with Russia's pursuit of Chechen rebels seeking shelter across the border in Georgia (mainly in the Pankisi Gorge).

have been under Moscow's control.[64] The 2014 annexation of Crimea, the precursor to the war in the Donbas, was in part a function of Russia's desire to control the naval base at Sevastopol; therefore, these earlier disputes plausibly could have escalated as events did in 2014.[65]

Table 3.4 summarizes whether factors are present or absent in the two disputes and in the 2014 conflict. The first column is identical to the respective column in Table 3.2; it is replicated here for ease of comparison.

Factors Unique to the Conflict

As Table 3.4 demonstrates, there are two factors present as drivers of escalation in the 2014 Ukraine conflict that are absent in the two disputes. First, in 2014, Russia perceived increased external security threats, generated by Ukraine's Maidan revolution and the country's resulting potential pro-Western turn, including potential NATO membership. Second, Moscow was overtly dissatisfied with its place in the international system. When the two disputes occurred in the early 1990s, neither NATO nor EU membership were realistic options for Kyiv, and Ukraine itself did not pose any external security threat to Russia. As already noted, we consider 2007 as the year after which Russia's dissatisfaction with its place in the world became an operational factor in its external behavior. The two disputes, by contrast, occurred at a time when Russia was still trying to understand its role in the new post–Cold War order. In those years, Moscow was often too consumed by its own domestic transformation and associated political and economic instability to be actively dissatisfied with its place in the international system.[66] However, by 2013–2014, Russian dissatisfaction with the international status quo was palpable. By then, repeated public statements by various Russian leaders underscored Moscow's preferences for a multipolar or polycen-

---

[64] N. A. Kryukov, "Osobennosti razvitiya i sostoyaniya rossiisko-ukrainskikh otnoshenii po pravovomu statusu Chernomorskogo flota RF," *Voennaya mysl'*, No. 5, 2006; Felgenhauer, 1999.

[65] Igor Delanoe, *Russia's Black Sea Fleet: Toward a Multiregional Force*, Arlington, Va.: Center for Naval Analyses, June 2019, p. 1.

[66] Allen C. Lynch, "The Evolution of Russian Foreign Policy in the 1990s," *Journal of Post-Communist Studies and Transition Politics*, Vol. 18, No. 1, 2002.

TABLE 3.4

Presence of Potential Drivers of Escalation in the 2014 Ukraine Conflict and Earlier Disputes

Potential Driver of Escalation	2014 Ukraine Conflict	Dispute 1: 1992 Division of Control over the Black Sea Fleet	Dispute 9: 1994 Cheleken Incident
Is this a territorial issue for Russia?	N	N	N
Is this a territorial issue for the opposing state?	Y	Y	Y
Is the disputed territory salient to Russia?	Y	Y	Y
Is the dispute or conflict located in a state hosting a Russian military installation?	Y[a]	Y	Y
Is the dispute or conflict located on or adjacent to territory of former Soviet Union?	Y	Y	Y
Does the opposing state share a land border with Russia?	Y	Y	Y
Is Russia experiencing acute increased uncertainty about the future?	Y	Y	Y
Does Russia face or perceive new or increased external security threats?	**Y**	**N**	**N**
Does Russia face domestic political instability?	**N**	**Y**	**Y**
Is the opposing state experiencing internal instability?	Y	Y	Y
Were there reputational costs for Russia for nonintervention?	Y	Y	Y
Was Russia the victor in a previous dispute with the opposing state?	N	N/A	N
Is power preponderance within the dyad in Russia's favor?	Y	Y	Y
Is Russia dissatisfied with its place in the international system?	**Y**	**N**	**N**
Is the opposing state a democracy?	Y	Y	N
Are Russian compatriots present on the territory where the conflict takes place?	Y	Y	Y

NOTE: Y= factor is present; N = factor is absent; N/A = factor not applicable. Bold text indicates drivers that are unique to the conflict.

[a] Russia had a military facility in Crimea, internationally recognized as Ukrainian territory at the time of the conflict.

tric international system as an alternative to U.S.-led unipolarity.[67] In his speech announcing the annexation of Crimea two months prior to the start of the Donbas conflict, President Putin made clear that Russia's actions were intended to push back on perceived U.S. attempts at global predominance.[68]

There is one factor unique to the disputes that is absent in the 2014 conflict. In the early 1990s, Russia experienced significant domestic instability, featuring a massive economic contraction and a constitutional crisis that culminated in the shelling of the parliament. There was no such instability in Russia in 2014. It is possible that the domestic instability of the 1990s acted as an "inhibitor" of escalation in the context of the two disputes while the authoritarian consolidation and economic prosperity of 2014 enabled the Kremlin to escalate on the ground in Ukraine without concern about domestic blowback.

Georgia Cases

The events underpinning the four disputes with Georgia all relate to the protracted conflicts over the breakaway regions of South Ossetia and Abkhazia. The disputes ranged from the alleged deployment of Russian military equipment in support of South Ossetian militia in late 2002 and early 2003 (dispute 39) to the September 2003 kidnapping of a Russian peacekeeper posted in Abkhazia (dispute 42). In the spring and summer of 2004, there was a series of incidents between Russia and Georgia relating to South Ossetia (dispute 43). Tensions flared again in 2005–2006, when there were several episodes involving Russian peacekeeping forces in Abkhazia and South Ossetia (dispute 44). See detailed narratives in Appendix F.

Factors Unique to the Conflict

As Table 3.5 demonstrates, three key drivers of the 2008 war are not present in the four disputes: Russia's dissatisfaction with the status quo in the international system; acute uncertainty about the future; and its perceptions of increased external security threats. For all three variables, there were rather

---

[67] See Elena Chebankova, "Russia's Idea of the Multipolar World Order: Origins and Main Dimensions," *Post-Soviet Affairs*, Vol. 33, No. 3, 2017.

[68] President of Russia, "Address by the President of the Russian Federation," speech, Moscow: The Kremlin, March 18, 2014a.

TABLE 3.5

Presence of Potential Drivers of Escalation in the Georgia Conflict and Earlier Disputes

Potential Driver of Escalation	2008 War	Dispute 39: 2003 Deployment of Russian Troops and Equipment to Tskhinvali	Dispute 42: 2003 Abduction of Russian Peacekeeper	Dispute 43: 2004 Increase in Tensions Between Russia and Georgia	Dispute 44: March 2005– September 2006 Simmering Tensions
Is this a territorial issue for Russia?	N	N	N	N	N
Is this a territorial issue for the opposing state?	Y	Y	Y	Y	Y
Is the disputed territory salient to Russia?	Y	Y	Y	Y	Y
Is the dispute or conflict located in a state hosting a Russian military installation?	Y	Y	Y	Y	Y
Is the dispute or conflict located on or adjacent to territory of former Soviet Union?	Y	Y	Y	Y	Y
Does the opposing state share a land border with Russia?	Y	Y	Y	Y	Y
Is Russia experiencing acute increased uncertainty about the future?	Y	N	N	N	N
Does Russia face or perceive new or increased external security threats?	Y	N	N	N	N

Table 3.5—Continued

Potential Driver of Escalation	2008 War	Dispute 39: 2003 Deployment of Russian Troops and Equipment to Tskhinvali	Dispute 42: 2003 Abduction of Russian Peacekeeper	Dispute 43: 2004 Increase in Tensions Between Russia and Georgia	Dispute 44: March 2005–September 2006 Simmering Tensions
Does Russia face domestic political instability?	N	N	N	N	N
Is the opposing state experiencing internal instability?	N	N	N	N	N
Were there reputational costs for Russia for nonintervention?	Y	Y	Y	Y	Y
Was Russia the victor in a previous dispute with the opposing state?	N	N	N	N	N
Is power preponderance within the dyad in Russia's favor?	Y	Y	Y	Y	Y
Is Russia dissatisfied with its place in the international system?	**Y**	**N**	**N**	**N**	**N**
Is the opposing state a democracy?	Y	Y	Y	Y	Y
Are Russian compatriots present on the territory where the conflict takes place?	Y	Y	Y	Y	Y

NOTE: Y= factor is present; N = factor is absent; N/A = factor not applicable. Bold text indicates drivers that are unique to the conflict.

marked differences between the circumstances under which the four disputes occurred and the 2008 escalation that resulted in the Russia-Georgia War.

The disputes all took place before Putin's 2007 Munich speech, the marker we use for the period after which Russia's discontent with its place in the international system was a factor in its external behavior. This discontent was heightened in the run-up to the 2008 conflict, particularly following the Western recognition of Kosovo's independence in February of that year. The four disputes with Georgia that did not escalate to conflict took place in a period when Russian discontent was not present as a potential driver.

The Bucharest NATO summit in April 2008, when members of the Alliance declared that Ukraine and Georgia "will become members of NATO," aggravated Russian uncertainty about the future and intensified perceptions of an imminent external threat.[69] By contrast, during the period when the four disputes unfolded (2003–2006), the prospect of a future dramatic shift in the regional balance of power—i.e., Georgia's NATO membership— seemed remote. And Russian concerns regarding external threats emanating from Georgia were not acute; any challenge to the Russia-backed separatists from Tbilisi was easily rebuffed, and Moscow did not need to worry about the possibility of NATO joining the fray.

The comparison of disputes with Georgia that did not escalate to conflict with the one that did, therefore, underscores the importance of two of the geopolitical factors in driving Russian flashpoints: uncertainty about the future and dissatisfaction with the status quo in the international system. However, as with the comparison of the Ukraine disputes and conflict, Moscow's perception of heightened external threats also played a key role.

Summary of Dispute Analysis

For both practical and methodological reasons, we did not conduct a detailed qualitative analysis of the drivers of onset for all 54 cases of Rus-

---

[69] Samuel Charap and Timothy J. Colton, *Everyone Loses: The Ukraine Crisis and the Ruinous Contest for Post-Soviet Eurasia*, New York: Routledge, January 2017, p. 88; NATO, "Bucharest Summit Declaration: Issued by the Heads of State and Government Participating in the Meeting of the North Atlantic Council in Bucharest on 3 April 2008," webpage, May 8, 2014; and Serhii Plokhy and M. E. Sarotte, "The Shoals of Ukraine: Where American Illusions and Great-Power Politics Collide," *Foreign Affairs*, January/February 2020.

sian disputes. However, we were able to compare disputes between Russia and two states with which it later had conflicts: Ukraine and Georgia. Two factors from our framework are present in conflicts with both countries that are absent in all the disputes: Russia's dissatisfaction with the international system and Moscow's perception of heightened external threats. In the Georgia cases, the additional factor of uncertainty about the future is present in the conflict but absent in the disputes.

Key Findings

In this chapter, we described how we developed a 16-factor framework of possible drivers of disputes and conflicts using the political science literature and then applied it to a variety of observations from our data set. The results of using the framework on detailed studies of four of the conflict cases (summarized in this chapter and provided in full in Appendixes B–E) highlight the importance of three categories of factors in driving escalation:

- **proximity:** The case studies indicate a causal relationship between proximity to Russia and escalation.
- **geopolitical drivers:** Russia's dissatisfaction with its place in the international system, acute uncertainty about the future, and reputational costs for nonintervention all played central roles when they were present.
- **external threat:** An immediate trigger for a flashpoint in all cases was Moscow's perception of a new or heightened external threat to security.

Applying the same framework to test for the presence of these 16 potential drivers in cases of disputes with Ukraine and Georgia that did not escalate to conflict underscored the importance of three factors from this initial list: dissatisfaction with the international systemic status quo, external threats, and acute uncertainty about the future. These factors are present when disputes with the same countries over roughly similar issues escalated to conflict; they are absent when those disputes did not escalate to conflict, suggesting a strong causal role.[70]

---

[70] Uncertainty about the future is absent only for the disputes with Georgia.

CHAPTER FOUR

Quantitative Analysis of Militarized Interstate Dispute Onset

This chapter uses statistical analytical techniques to identify the correlates or variables most closely associated with the onset of MIDs involving Russia. The data set described in Chapter Two contains only 63 total observations (54 disputes and nine conflicts) and is thus not suitable for such a quantitative analysis. Therefore, we analyzed the dyadic MIDs data set, which uses country-year pairs as the unit of analysis, thus dramatically increasing the sample size for our statistical analyses.[1] This structure also allows for meaningful variation in the dependent variable; that is, the outcomes we examine provide both MID onset and non-onset.

As discussed in the previous chapter, the literature on conflict and dispute onset and on escalation suggests several factors that might play a primary role in dispute propensity. Here we assess which of these variables have predictive power, from a statistical point of view, on MIDs with Russia.

---

[1] We therefore cannot make the same distinction between militarized disputes and militarized conflicts made in Chapters Two and Three. Our dependent variable in this chapter is a (country-year pair) MID, which encompasses both disputes and conflicts and excludes interventions in civil conflicts on behalf of sitting governments. The dyadic data set also does not allow us to distinguish between cases in which Russia was a primary actor and those in which it played a secondary role. We did attempt to differentiate among Russia's MIDs by level of hostility. (MIDs are ranked by a 5-point hostility scale.) However, the limitations of the data prevented such an inquiry. More specifically, our sample size—particularly the scarcity of MIDs involving Russia during the sample period—precluded us from precisely identifying the factors associated with relatively higher-hostility MIDs. Preliminary analytical attempts using an ordered logit model revealed that a significant proportion (nearly half) of our observations were completely determined, making our standard errors unreliable.

To the extent possible, we attempted to complement the qualitative analysis in the previous chapter by testing the explanatory power of similar variables. To be clear, the models we employ in this chapter do not evaluate the possible causation associated with each variable; rather, they assess the predictive potential of each on MID onset.[2]

Drawing on this first line of effort, we aim to better understand which states are most likely to experience a future dispute with Russia premised on historical data and respective country attributes. Put differently, having identified possible factors that correlate with Russian MID onset, can we then say which countries might be at most risk of experiencing a militarized run-in with Russia? Again, we seek to do so through data analysis and predictive statistical estimation.

Method and Research Design

This chapter presents two primary strands of quantitative analysis. The first is a series of statistical models designed to estimate MID onset between Russia and 169 other countries, employing data from 1992 to 2010. These models allow us to establish baseline estimations of the coefficients of the key correlates associated with Russian MID onset during this 19-year span. A second set of analysis incorporates the statistical results from the first modeling effort to predict the onset of Russian MIDs using data for 2011 to 2018.

We designed this research approach in line with the goals of the project described in Chapter One and mindful of the data available to us. The MIDs data, produced and released by the Correlates of War Project, extended only through 2010 at the time that this research was completed. We therefore lack information about our dependent variable (MID onset) after this year. However, our data on the key independent variables that we tested are not similarly limited; they run through 2018. We therefore leveraged the model (coefficient) estimates incorporating the 1992–2010 MIDs data to conduct predictive analysis that takes advantage of data on our independent vari-

---

[2]   To reiterate, for this analysis we do not alter or vary the definition of a MID from that offered by the Correlates of War Project (Jones, Bremer, and Singer, 1996). All references in this chapter to *dispute* refer to the standard Correlates of War definition of MID.

ables from 2011 to 2018. We did this for two reasons. First, we would like to draw a distinction between the sample used to estimate the regression coefficients in the baseline model and the sample used for the predictive analysis that follows. If we were to use the same sample for both components of the analysis, then the predictive aspect would tend toward replicating the MIDs that occurred during that sample period. Second, we would like our predictions to apply to as recent a time frame as possible for the sake of relevance to the current environment. For this reason we used data on our explanatory variables up though the year 2018.

We employed Firth logit models, similar in design to logit and rare events logit models, to estimate MIDs.[3] To do so, we employed yearly, nondirected, dyadic analysis.[4] The Firth technique is specifically designed for data sets in which observations are infrequent. This tool is useful for our data set because there are relatively few MIDs between Russia and other states relative to the total number of dyad-year observations for all countries. For these models, we did not refine or alter the MIDs data set in the manner described in Chapter Two.

---

[3] The results we produce are robust across all three techniques, with only minor estimation differences. On these techniques, see David Firth, "Bias Reduction of Maximum Likelihood Estimates," *Biometrika*, Vol. 80, No. 1, March 1993; and Gary King and Langche Zeng, "Logistic Regression in Rare Events Data," *Political Analysis*, Vol. 9, No. 2, 2001.

[4] There is a debate concerning the use of directed versus nondirected dyads to estimate conflict or dispute onset. In general, a nondirected approach treats, for example, a Russia-Japan observation in a given year the same as it treats a Japan-Russia observation of the same year—there is no distinction between the two. Hence, there is only one observation per dyad-year-MID in nondirected dyadic analyses. A directed-dyad analysis allows for behavioral choices and dyadic outcomes to be different in the two directions, thereby permitting simultaneous testing of different hypotheses. Continuing with our earlier example, in directed-dyad analyses, Russia-Japan and Japan-Russia observations would both appear separately in the data set. We use nondirected dyads in our analysis because we are concerned with analyzing factors related to MIDs with Russia irrespective of which side initiated the dispute. Moreover, none of our explanatory variables depend on the direction of the dyad; thus, if we were to use directed dyads, we would in essence be artificially doubling the size of our data set. For more on this, see D. Scott Bennett and Allan C. Stam, "Research Design and Estimator Choices in the Analysis of Interstate Dyads: When Decisions Matter," *Journal of Conflict Resolution*, Vol. 44, No. 5, 2000.

As noted in Chapter Three, there is a rich literature on dispute and conflict onset. We used this literature to inform our decisions about the explanatory variables we tested in the models. Ultimately we chose to test the following variables:

- Because conflict is often related to competing territorial or maritime claims, we coded a dummy variable capturing a *territorial or maritime disagreement* between Russia and the opposing state.[5] We consulted three sources to code this variable. First, we obtained data on territorial disagreements involving Russia (through 2013) from the Issue Correlates of War Territorial Claims Dataset. We supplemented this with data on maritime disagreements involving Russia (through 2016) from the Issue Correlates of War Maritime Claims Dataset. Finally, we refined this data using input from subject-matter experts, and extended it up to 2018.[6]
- The models test the significance of Russia's relative power vis-à-vis the opposing state in the MID. We measured power using a component of the Global Power Index (GPI), which is synthesized from subindexes of military, political, economic, technological, and demo-

---

[5] We also include Russia's riverine disagreement with China.

[6] Our final list of states engaged in territorial or maritime disputes with Russia is as follows: Azerbaijan (1991–2001), China (1954–2004), Estonia (1991–ongoing), Georgia (1996–ongoing), Iceland (1998–1999), Iran (1991–2019), Japan (1952–ongoing), Kazakhstan (1991–2005), Latvia (1991–2007), Lithuania (1991–ongoing), Norway (1970–2010), Poland (1991–2001), Turkey (1994–1998), Ukraine (1993–ongoing), and the United States (1900–ongoing). Paul R. Hensel, "Contentious Issues and World Politics: The Management of Territorial Claims in the Americas, 1816–1992," *International Studies Quarterly*, Vol. 45, No. 1, March 2001; Paul R. Hensel and Bryan Frederick, "Provisional Issue Correlates of War Territorial Claims Dataset," version 1.02, unpublished manuscript, January 1, 2017; Paul R. Hensel, Sara McLaughlin Mitchell, and Thomas E. Sowers II, "Conflict Management of Riparian Disputes: A Regional Comparison of Dispute Resolution," *Political Geography*, Vol. 25, No. 4, May 2006; and Paul R. Hensel, Sara McLaughlin Mitchell, Thomas E. Sowers II, and Clayton L. Thyne, "Bones of Contention: Comparing Territorial, Maritime, and River Issues," *Journal of Conflict Resolution*, Vol. 52, No. 1, February 2008.

graphic power.[7] We specifically used the GPI military subindex in the models in this chapter.

- We examined the possible effect of Russian state capacity on its ability to engage in militarized disputes. Specifically, we incorporated a variable reflecting Russian real oil rents (in 2010 U.S. dollars) per capita; *oil rents* are defined as the difference between the economic value of crude oil production (i.e., value of oil production at world prices) and total production costs. This variable comes from the World Bank World Development Indicators data base.[8]

- We also created a dummy variable indicating whether a country was formerly a Soviet republic. According to the history of Russian military activities in post-Soviet Eurasia, this is an important potential explanatory factor of Russian MIDs.

- To examine the significance of geographic proximity of the opposing state to Russia for MID onset, we created a dummy variable for terri-

---

[7] Jonathan D. Moyer and Alanna Markle, *Relative National Power Codebook*, Denver, Colo.: Frederick S. Pardee Center for International Futures, Josef Korbel School of International Studies, University of Denver, version 7.2.2018, 2017. Although it is traditional to use scores from the Composite Index of National Capability (CINC) to operationalize power, we do not do so here. The CINC indicator of relative power is based on such state characteristics as population, urban population, iron and steel production, energy consumption, military expenditure, and military size. We elected to use the alternative GPI measure in our baseline specification because it better captures the nature of power in the postindustrial age. The GPI is constructed from subindexes of military, political, economic, technological, and demographic power (each of which, in turn, is constructed from more-precise indicators). We used the military subindex, which is composed of a country's (1) share of global military personnel, (2) share of global military spending, and (3) share of global nuclear weapons (logged). The GPI was developed on behalf of the National Intelligence Council by the Frederick S. Pardee Center for International Futures at the University of Denver. For additional discussion of the merits of using GPI instead of CINC, see Jacob L. Heim and Benjamin M. Miller, *Measuring Power, Power Cycles, and the Risk of Great-Power War in the 21st Century*, Santa Monica, Calif.: RAND Corporation, RR-2989-RC, 2020.

[8] World Bank, World Development Indicators, data catalog, undated-b. At least one effort has found that aggressive foreign policy rhetoric in Russian presidential speeches positively correlates to oil prices. See Maria Snegovaya, "What Factors Contribute to the Aggressive Foreign Policy of Russian Leaders?" *Problems of Post-Communism*, Vol. 67, No. 1, 2020.

torial *contiguity* with Russia. This variable captures only shared land borders, not maritime boundaries.[9]

- To test for the significance of regime type in determining MID onset, we operationalize the level of democracy exhibited by Russia and its opponents using Polity IV scores.[10]

- We also used an interaction term of the Polity IV scores of Russia and its opponent to examine how the two jointly determine the likelihood of MID onset. Because the relationship between Russia's level of democracy and likelihood of a MID with another state might be a function of the other state's level of democracy, or the effect of the opponent state's level of democracy on the likelihood of a MID with Russia might be a function of Russia's level of democracy, this additional variable was necessary.[11]

- The academic literature on defense alliances suggests that such partnerships might serve as both bulwarks against war and potential augers of dispute. Therefore, we coded a defense pact variable indicating the presence of a formal defense pact or alliance obligation between Russia and the opposing state.[12]

---

[9] The contiguity and territorial and maritime disagreement variables share a considerable amount of overlap; i.e., they take on the same value (either 0 or 1) in the vast majority of observations. Even though there is a substantial degree of overlap between the two variables, we used both in our model specifications. This permitted us to test for effects of territorial disputes on MID onset with nations other than those with which Russia shares a land border. For further discussion, see Braithwaite and Lemke, 2011.

[10] Polity IV scores range from –10 (lowest level of democracy) to +10 (highest level). It is worth noting that there is appreciable variation in Russia's Polity IV score from 1992 to 2010. Specifically, Russia's Polity IV score varies from 3 to 6 over our sample period.

[11] In our statistical analysis, we add 11 to each Polity IV score so that the variable ranges from 1 to 21. Doing so prevents issues with the interaction term that can arise when one or both scores are negative or 0 values. Monty G. Marshall, Ted Robert Gurr, and Keith Jaggers, *Polity IV Project: Political Regime Characteristics and Transitions, 1800–2018: Dataset Users' Manual*, Vienna, Va.: Center for Systemic Peace, July 27, 2019.

[12] We originally sourced this variable from the Alliance Treaty Obligations and Provisions data set but had to make considerable changes to it. We discovered numerous errors concerning a lack of guarantee of mutual assistance or inclusion of Soviet-era agreements that were being dismantled in 1992. The Russian alliances we coded were Armenia (1994–ongoing), Azerbaijan (1994–1999), Belarus (1994–ongoing), Geor-

- We also created a variable reflecting the *foreign policy disagreement* between Russia and the opposing state. The foreign policy disagreement variable is constructed from United Nations General Assembly voting data using a method applied by Bailey et al.[13] Higher values in the index correspond to greater disagreements between Russia and other states.[14]
- To examine the possibility that Russia's propensity to become involved in disputes with major powers is distinct from other countries, we used a dummy variable reflecting the *major power* status of opponents. We coded the United States, United Kingdom, France, Germany, China, and Japan as major powers for this purpose.[15]
- The number of years that have passed since a given state has had a MID with Russia might affect the propensity of future MIDs between the two states. To account for this possibility, we incorporated a time

---

gia (1994–1999), Kyrgyzstan (1994–ongoing), Kazakhstan (1994–ongoing), Tajikistan (1994–ongoing), and Uzbekistan (1994–1999; 2006–2012). On the various nuances of alliance commitment and their respective impacts on conflict deterrence, see Brett Ashley Leeds, "Do Alliances Deter Aggression? The Influence of Military Alliances on the Initiation of Militarized Interstate Disputes," *American Journal of Political Science*, Vol. 47, No. 3, July 2003. On increased conflict propensity of allies, see Brett V. Benson, "Unpacking Alliances: Deterrent and Compellent Alliances and Their Relationship with Conflict, 1816–2000," *Journal of Politics*, Vol. 73, No. 4, 2011; and James L. Ray, "Friends as Foes: International Conflict and Wars Between Formal Allies," in Charles S. Gochman and Alan Ned Sabrosky, eds., *Prisoners of War? Nation-States in the Modern Era*, Lexington, Mass.: Lexington Books, May 1, 1990. For a theoretical treatment of alliance between rivals, see Emerson M. S. Niou and Sean M. Zeigler, "External Threat, Internal Rivalry, and Alliance Formation," *Journal of Politics*, Vol. 81, No. 2, April 2019.

[13] Michael A. Bailey, Anton Strezhnev, and Erik Voeten, "Estimating Dynamic State Preferences from United Nations Voting Data," *Journal of Conflict Resolution*, Vol. 61, No. 2, 2017.

[14] More precisely, the index incorporates United Nations (UN) voting data to calculate each country's ideal point on a single dimension reflecting its views of the U.S.-led liberal order. Foreign policy disagreement between states A and B is defined as the absolute difference between those two countries' ideal points. This alternative to simply measuring the similarity of states' preferences based solely on UN voting similarity is likely preferable because UN voting similarity in a given year is affected by the mix of issues under consideration that year. In our data set, the index values range from 0.000 to 3.222 (Bailey et al., 2017).

[15] Correlates of War Project, "State System Membership List," ver. 2016, 2017.

control. This variable, *years since previous MID*, begins in 1991 and reflects a linear relationship between time elapsed since a previous MID and MID propensity.[16] We drop observations corresponding to ongoing MIDs—i.e., for a given MID, we exclude observations subsequent to the initial year of dispute.[17]

Descriptive Statistics

Table 4.1 offers summary statistics of our variables in (yearly) dyadic (pairs of states) format. This table lists the number of observations (in dyadic form), the mean, standard deviation, and minimum and maximum values for each variable. To gain a better appreciation for how the dependent variable varies over time, Table 4.2 shows the number of yearly MID onsets with Russia from 1992 to 2010. The table also lists the country or countries party to each MID.

Modeling Results

Table 4.3 displays regression coefficient estimates and standard errors from four separate models. First, we discuss the technical results. Then, we consider what the results mean in practical terms. Because our contiguity measure, our territorial disagreement indicator, and our binary variable for former Soviet states exhibited a fair amount of overlap, we initially treated

---

[16] In alternative specifications, we used a squared and cubed version of this control to capture potentially nonlinear temporal effects. For further discussion, see David B. Carter and Curtis S. Signorino, "Back to the Future: Modeling Time Dependence in Binary Data," *Political Analysis*, Vol. 18, No. 3, 2010. The higher-order controls had estimated coefficients near zero in magnitude that were not statistically significant at the 5-percent level, and their inclusion did not significantly alter the estimated effects associated with other regressors. We ultimately did not use these controls in our primary specification for these reasons and also because of the limited number of observations and rare events that characterize our data set.

[17] See Bennett and Stam, 2000, for further discussion on this technique.

TABLE 4.1

Summary Statistics (Dyadic)

Variable	Observations	Mean	Standard Deviation	Minimum	Maximum
MID onset with Russia	3,587	0.022	0.146	0.000	1.000
Territorial or maritime disagreement	3,587	0.055	0.229	0.000	1.000
Contiguity	3,587	0.072	0.259	0.000	1.000
Military power of Russia	3,587	7.551	0.676	6.705	9.237
Military power of opposing state	3,181	0.572	2.808	0.000	36.112
Real oil rents per capita	3,585	690.87	383.1645	90.8615	1283.028
Military power ratio (Russia/opposing state)	3,103	3,947.161	24,877.970	0.196	657,573.200
Russia Polity IV democracy score	3,587	15.432	1.311	14.000	17.000
Opposing state Polity IV democracy score	3,041	14.024	6.657	1.000	21.000
Polity IV democracy interaction	3,041	216.629	105.206	14.000	357.000
Defense pact	3,587	0.030	0.171	0.000	1.000
Foreign policy disagreement	3,466	0.917	0.594	0.001	3.222
Major power (opposing state)	3,587	0.032	0.177	0.000	1.000
Former Soviet republic (opposing state)	3,587	0.078	0.267	0.000	1.000
Time since last MID	3,587	41.271	38.539	0.000	194.000

TABLE 4.2

Number of Militarized Interstate Dispute Onsets Involving Russia, 1992–2010

Year	Number of MID Onsets	Countries Involved
1992	5	Estonia, Georgia, Moldova, Sweden, Ukraine
1993	8	Iraq, Azerbaijan, Turkey, Japan, Moldova, China, Afghanistan, Poland
1994	4	Ukraine, Afghanistan, Latvia, China
1995	1	Lithuania
1996	3	Ukraine, Turkey, Japan
1997	4	Georgia, Poland, United States, Afghanistan
1998	4	Latvia, Serbia, Afghanistan, Norway
1999	7	Afghanistan, Georgia, United Kingdom, Norway, Azerbaijan, Turkey, [Belgium, Canada, Denmark, France, Germany, Italy, Norway, Netherlands, Portugal, Spain, Turkey, United Kingdom, United States]
2000	6	[Canada, United States], United States (2x), Turkey, Japan, Serbia
2001	4	Afghanistan, Georgia, Japan, Norway
2002	2	Argentina, Azerbaijan
2003	4	Denmark, Georgia (2x), Sudan
2004	1	Georgia
2005	3	Georgia, Norway, Ukraine
2006	2	Japan (2x)
2007	2	Finland, Georgia
2008	3	Japan, Norway, Ukraine
2009	2	China, Japan
2010	0	
Total:	65	

NOTE: Brackets [ ] denote cases where multiple countries were involved in a single MID onset with Russia.

TABLE 4.3

Firth Logit Regressions—Linear Time Controls

Variable	Model 1 (contiguity)	Model 2 (disagreement)	Model 3 (former Soviet)	Model 4 (baseline)
Contiguity	1.68442*** (0.29080)			1.04666*** (0.39908)
Territorial or maritime disagreement		1.77454*** (0.30340)		1.02099** (0.39960)
Former Soviet republic			1.51496*** (0.33707)	
Real oil rents (Russia)	0.00163*** (0.00051)	0.00164*** (0.00054)	0.00192*** (0.00051)	0.00149*** (0.00053)
Military power of opposing state	0.04365* (0.02534)	0.00885 (0.02665)	0.03306 (0.02521)	0.02156 (0.02690)
Military power ratio (Russia/opposing state)	0.00000 (0.00001)	-0.00000 (0.00001)	-0.00000 (0.00001)	0.00000 (0.00001)
Russia's Polity IV score	-0.51807 (0.35234)	-0.35438 (0.33880)	-0.50027 (0.35675)	-0.41830 (0.35221)
Opposing State's Polity IV score	-0.09813 (0.28436)	0.04058 (0.27831)	-0.04823 (0.28788)	-0.02317 (0.28721)
Polity IV interaction	0.00884 (0.01873)	0.00151 (0.01830)	0.00526 (0.01905)	0.00350 (0.01890)

Table 4.3—Continued

Variable	Model 1 (contiguity)	Model 2 (disagreement)	Model 3 (former Soviet)	Model 4 (baseline)
Defense pact	-0.16890 (0.66104)	-0.11910 (0.67795)	-0.54823 (0.66782)	-0.23277 (0.67319)
Foreign policy disagreement	-0.50061* (0.27093)	-0.47533* (0.26630)	-0.42567 (0.26473)	-0.43732 (0.26978)
Opposing state is a major power	1.47375*** (0.38614)	0.88813** (0.41960)	1.57051*** (0.41035)	1.23868*** (0.41288)
Time since MID	-0.27099*** (0.04240)	-0.25694*** (0.04353)	-0.29349*** (0.04293)	-0.24564*** (0.04317)
N	2872	2872	2872	2872

NOTE: Standard errors in parentheses. *, **, and *** indicate coefficients statistically different from zero at the 0.10, 0.05, and 0.01 levels, respectively.

these covariates in separate models.[18] Put differently, these indicators are highly, though not completely, correlated: By definition, the majority of Russia's territorial disagreements are with states on its borders, and many of these states were also part of the former Soviet Union. Therefore, as a point of departure, we kept these variables separated in our models to probe the potential explanatory power of each. Model 1 incorporates the contiguity measure; model 2 features the territorial or maritime disagreement indicator; and model 3 incorporates the former Soviet Republic variable. Several important and nuanced results emerge from the three models.

The first model indicates that contiguity is a statistically significant predictor of dispute onset. The coefficient is positive and significant at the 0.01 level. However, when we replace this measure with the territorial disagreement variable (model 2), it suggests that it too is a significant predictor of dispute onset at the 0.01 level. Finally, when we instead use the indicator for a former Soviet republic (model 3), we see that its estimated coefficient is also statistically significant at the same level. Taken together, the results from these models indicate that all three of these variables (contiguity, territorial disagreement, and former Soviet republic) have some explanatory power. The estimated coefficients for these three covariates are relatively close. The largest is the territorial disagreement variable at 1.77; the contiguity coefficient is 1.68; and the estimated coefficient for former Soviet republic is 1.51.

In an attempt to further unpack these findings, model 4 incorporates the measures for both contiguity and territorial or maritime disagreements but excludes the variable of being a former Soviet republic.[19] We used this as our baseline model and employed it for our predictive analysis (which we describe in a later section of this chapter). The model's estimated results

---

[18] These variables exhibit collinearity (though not perfect collinearity). This is potentially problematic because when variables in the same regression model are correlated, they cannot independently predict the value of the dependent variable.

[19] For this model's specification, we used two of the three related variables because of their high correlation and because we wanted as parsimonious a model as possible, given our limited data (both in terms of total observations and number of MIDs) so as to not overfit the model. When we did use all three variables in the same model (not shown), the statistical procedure was unsure to which variable it should attribute effects—standard errors increased precipitously.

indicate that both factors likely play a part in Russian MID onset. The estimated coefficient for the contiguity variable is again positive and significant at the 0.01 level. The territorial or maritime covariate is also a significant predictor of dispute onset but at the 0.05 level. Both coefficients fall slightly (from model 2). These results suggest that although both factors are important predictors of Russian MID onset, unpacking their precise effects is challenging.[20]

The four models make clear a few key results. First, indicators of territorial claims, contiguity with Russia, and being a former Soviet republic all reflect similar statistical information and appear to have explanatory power. However, the coefficient estimates for these variables change depending on which of them we use in our models. That is, the results of the models are sensitive to the inclusion of correlated variables. This is because there is considerable conceptual overlap across the three variables. Of the three, the largest estimated coefficient is for the territorial disagreement variable.

A few other modeling results are also worth highlighting. The major power status of opposing countries is a positive and statistically significant correlate of disputes with Russia. This result is consistent across all four models. It is driven by disputes with China (three), France (one), Germany (one), Japan (eight), the United Kingdom (two), and the United States (five). It bears noting that our dependent variable in these models is the onset of a MID, not militarized conflict initiation (as defined in Chapter Two).[21] Another factor consistently correlated (negatively) with Russian disputes

---

[20] In an effort to untangle these relationships a bit more, we estimated another model (not shown) that separated the contiguity variable. We created a Russian contiguity variable that *excluded* countries that were part of the former Soviet Union. Along with this revised contiguity measure, the model used the territorial disagreements variable and an indicator of former Soviet states (along with the other covariates). The results of this model indicated (in line with the results we have already described) that all three variables are statistically significant predictors of MID onset. The former Soviet republic coefficient was significant at the 0.10 level; the territorial disagreement variable was significant at the 0.01 level; and the (revised) contiguity variable was significant at the 0.05 level. This result offers further confidence that all of these variables have substantive predictive impacts on dispute onset.

[21] A paucity of conflict observations prevented this kind of analytical effort. We describe qualitative escalation analysis in Chapter Three. To be clear, some of the MIDs reached conflict levels, but most did not.

is *time*. Each year that passes after a dispute onset with Russia lowers the likelihood of a future dispute. The estimated effect is not large but is statistically significant across all four of the models. The oil variable is also significantly correlated with disputes, but the estimated coefficients for this variable in the four models are not large, especially relative to the other significant coefficient estimates.

None of the other correlates that we used in the model demonstrates any consistently significant effects across the four models. Although the military power ratio coefficient in model 1 is statistically significant at the 0.10 level, the estimated coefficient is small.[22] Finally, the estimated coefficient for the foreign policy disagreement variable is negative and statistically significant at the 0.10 level in models 1 and 2. The coefficients for this indicator in the other two models are also negative, although not statistically significant. This tentatively suggests that foreign policy disagreements with Russia might lower the chances of a MID with Russia, if slightly. But the results are inconsistent.[23]

To gain a better sense of the substantive relationships between key variables and Russian MID onset we also estimate how the expected probability

---

[22] It is worth noting that we investigated several manners of quantifying power. The baseline model uses the GPI *military power* variable. However, there is little substantive change when we use the GPI *total power* measure. As an alternative, we also used CINC scores. Doing so produced fairly similar estimates on the whole, albeit with some differences. For example, when we used CINC scores, the estimated size of the effect of other-state power—while operating in the same direction—was larger in magnitude than when we used GPI power. In either case, the coefficients associated with other-state power are imprecisely estimated. Also, the power ratio variable is statistically significant in the specification employing CINC scores to measure power. Although statistically significant, the estimated effect is very near 0—in line with the GPI estimates.

[23] We further conducted several robustness checks and made alternative specifications to some variables, none of which produced any substantive changes to the findings we present here. These tests consisted of coding Russia as either an anocracy or democracy based on its Polity IV score. We also tested additional specifications related to the policy disagreement variable. First, we added an interaction term between contiguity and the policy disagreement variable. Second, we replaced contiguity with the distance between capital cities and added an interaction term between the distance and policy disagreement variables. Third, we replaced contiguity with the distance between country borders and added an interaction term between this variable and the policy disagreement variable. None of these explorations were informative.

of a militarized dispute changes as we vary the key factor covariate. To do so, we used a stepwise process for each of the following variables (whose coefficients are statistically significant in our baseline model): *contiguity, territorial disagreement, real oil rents, major power,* and *time since MID*.[24] First, holding each variable of interest at its mean, we estimated predicted probabilities of MID onset for the countries in the data. We then averaged these country-specific predictions to generate a baseline estimate, again with the variable of interest held to its sample mean. The second step involved repeating the first step but increasing each independent variable of interest by two standard deviations above its sample mean for these calculations. Predicted probabilities were again estimated over the entire sample of countries.

Table 4.4 displays the results of this process. The first (far-left) column of the table lists the five variables of concern. The second column displays the predicted probability of a MID while holding the variable of interest at its sample mean. Again, these estimates were computed for each country and then averaged. The third column again reports (averaged) predicted probabilities of MID onset, but instead uses values of two standard deviations

TABLE 4.4

Change in Probability of Militarized Interstate Disputes Involving Russia by Key Factor

Variable	Average Probability (mean)	Average Probability (mean + 2 SD)	% Change in Probability
Contiguity	1.750%	2.940%	68.016
Territorial or maritime disagreement	1.750%	2.585%	47.718
Real oil rents (Russia) per capita	1.750%	3.566%	103.796
Opposing state is a major power	1.750%	2.798%	59.871
Time since MID	1.750%	0.651%	−62.793

NOTES: These are the predicted effects of a two–standard deviation increase in each listed variable on the average (across time and space) predicted probability of a MID involving Russia occurring. SD = standard deviation.

---

[24] All calculations use model 4, our baseline, from Table 4.3. Because this model does not use *former Soviet* estimates, we do not list that variable in our substantive effects calculations.

above the sample mean for each variable. The final (far-right) column displays the percentage change in predicted probabilities reported in the previous two columns.

A few points of interest emerge from Table 4.4. First, the baseline probability of a MID with Russia is quite low (1.75 percent). This is not unexpected, given the rarity of MIDs in the data set. Second, of the five primary variables of interest, the largest change in predicted probability (from a two–standard deviation increase above the sample mean) is associated with a change in the *oil rents* variable. Although the suggested effect is apparently large (over 100 percent increase in probability), this might be an artifice of the variable's distribution. A doubling of the mean of the *oil rents* variable puts it at just over its maximum value—an unlikely event. Still, an increase in oil revenue is associated with a strong jump in the probability of a dispute. The estimated effects for the other variables are more muted. For the *contiguity* indicator, the percentage change in predicted probability is a jump of over 9 percent. Doubling the sample mean for *major power* status and *territorial disagreements* is associated with an appreciably smaller increase in the predicted probability of a MID with Russia. Finally, doubling the average time since MID also lowers the percentage change in the probability of a militarized dispute. Although the percentage change is large (a fall of more than 60 percent), the effect in absolute terms is not very big.

Predictive Analysis

We used the results of the modeling described in this chapter to conduct predictive analysis. Our intent was to generate a list of countries that present the highest probability of experiencing a MID with Russia according to how well their respective characteristics and past history of interaction with Russia (e.g., territorial disagreements, shared borders, power) align with those factors identified in the models. To do so, we combined the estimated coefficients from our baseline model (model 4) with more-recent, country-specific (independent variable) data from 2011 to 2018. As noted, the baseline model used data from 1992 to 2010 to generate an estimated intercept and coefficients associated with each explanatory variable. For the predictive analysis, we used these intercepts and coefficients to generate predictions for all countries in the data set for each year between 2011 and 2018, using the data we had for each of the independent variables during this time frame.[25] These yearly predictions for each country were then averaged across years to produce a final country-specific value.[26]

This approach has several advantages and limitations. First we were able to generate our predictions on out-of-sample data, a generally desirable aspect of this type of analysis.[27] However, we were unable to test the model's accuracy because we lacked MIDs data after 2010. That is, we were unable to cross-check our model's predictive results because we had no data on MIDs covering the years in our prediction window. Our model is also limited by construction and by data: We were only able to use a fixed number of variables for which we had data. Other variables not captured in the model (such as geopolitical factors) might also have some predictive impact.[28] That said,

---

[25] The one "problem" independent variable is "time since MID." The MID data set ends in 2010; therefore, we lacked the ability to code this variable through 2018. To get around this, we computed the average time since MID for each country within the sample period (1992–2010). We then used this same value for the time since MID variable for a given country for each year between 2011 and 2018.

[26] Formally, the process is as follows. The first step is to obtain regression coefficients and an intercept from the baseline Firth logit model. Then, for a given year and a given country, these outputs (intercept and coefficients) are combined (i.e., multiplied) with the respective values of the associated covariates. Exponentiating this expression (the fitted right-hand side of the regression equation) produces what is known as *the odds*. The predicted probability is then computed as odds / (1+odds). The probability equation for a given country in a given year is given by

$$P = \frac{e^{\left(\alpha + \beta_1 X_1 + \ldots + \beta_n X_n\right)}}{1 + e^{\left(\alpha + \beta_1 X_1 + \ldots + \beta_n X_n\right)}} \, ,$$

where P denotes the predicted probability.

[27] For more on predictions, statistical significance in modeling, and limitations to this approach see Michael D. Ward, Brian D. Greenhill, and Kristin M. Bakke, "The Perils of Policy by P-Value: Predicting Civil Conflicts," *Journal of Peace Research*, Vol. 47, No. 4, 2010.

[28] It is worth noting that we discuss many of these factors in our qualitative analyses in other chapters of this report.

our model is able to capture both empirically and theoretically important factors highly correlated with Russian MID onset.

Table 4.5 displays the results of this process. The countries in the list are ordered from highest to lowest estimated probability of a MID onset with Russia. The table lists only the top 17 countries generated by the analysis.[29] Of note, the countries at the top of the list include the United States, several

TABLE 4.5

Country Predictions of a Militarized Interstate Dispute with Russia

Country	Estimated Probability
Japan	0.518
Ukraine	0.381
United States	0.327
China	0.209
Norway	0.195
Lithuania	0.168
Germany	0.124
Estonia	0.119
Azerbaijan	0.115
United Kingdom	0.111
Latvia	0.107
Poland	0.101
Montenegro	0.071
Finland	0.068
Turkey	0.061
Yugoslavia / Serbia	0.057

NOTE: Regression models used a linear time control. The table lists only the top 17 countries predicted by the analysis.

[29] The ordered list is remarkably consistent irrespective of which model we use for the predictive analysis. For instance, employing models 1, 2, 3, or 4 all generate identical orderings of the top six countries.

NATO members, and Japan (an important U.S. ally). The list also includes China. Although our motivation for this research was to look at possible flashpoint scenarios with non-NATO allies in Europe that could entangle the United States, these results underscore the potential for militarized engagements with Russia on a broad scale.

Several points are worth highlighting. First, the model suggests that Russian disputes with the United States itself are not unlikely. It bears repeating, however, that the model does not speak to the character or severity of such disputes: They could vary from a tussle over fishing rights in the Bering Strait to armed conflict. Second, the likelihood of a dispute between Russia and a U.S. ally is by no means negligible. Of the top 13 countries listed, nine are formal treaty allies of the United States. Again, past disputes have not escalated to a conflict that required U.S. forces to respond directly, and future disputes might not, either. Additionally, the results suggest that the chances for Sino-Russian disputes should not be discounted. Although China resolved its formal territorial disagreement with Russia in 2004, it is a major power sharing a large land border with Russia—two factors highly correlated with MID onset. Lastly, as already noted, we could not verify the accuracy of our predictive model because the MIDs data set had not been updated for the period from 2011 through 2018. However, observationally, several countries at the top of Table 4.4 have been militarily involved at various levels of intensity with Russia in that period: Ukraine, Norway, and the Baltic countries. This offers, at least anecdotally, validity to the modeling approach taken here.

Key Findings

This chapter provides a detailed look at Russian MID onset from a distinctly quantitative perspective. Several key findings emerge from the modeling efforts:

- States with unregulated land and maritime border issues with Russia are more likely to find themselves involved in a MID with Russia. This variable is a consistent and statistically significant predictor of a MID with Russia.

- Relatedly, contiguity with Russia is also predictive of a higher likelihood of dispute onset. A shared border with Russia is strongly correlated with dispute onset.
- Being a former Soviet republic also creates an increased probability of MIDs with Russia. Although the effect looks to be marginally less than that of states on Russia's border or those with territorial disagreements, it is nonetheless a significant explanatory factor.
- Although it is difficult to untangle with a high degree of confidence, the separate effects of these three variables all appear to matter in the models we present. However, the results are sensitive to which variables are specified within the models.
- Russian oil revenues are also positively correlated with Russian MID propensity.
- Another factor to emerge from the quantitative models associated with Russian MID onset with other states is time elapsed. Simply put, a recent militarized dispute between Russia and another state strongly suggests that another might soon be in the offing.
- Russia's opponents in MIDs are not limited to smaller or economically weaker states. Russia has a history of militarized disputes with major powers in the international system. This fact is clearly reflected in the statistical models.

It is worth reiterating that the suite of statistical models and analysis presented here specifically deals with MIDs. Therefore, the data (on MIDs) encompass both interstate clashes that escalate to the point of combat deaths and those that do not. As previously discussed, some have argued that the causes of militarized disputes, fatal disputes, and wars are quite similar.[30] Although the causes of militarized disputes and conflict might not dramatically differ, we have not explicitly drawn connections between our results on the correlates of Russian MIDs and possible correlates of Russian conflicts.

Finally, the models presented here speak to a variety of cases in which we can reasonably expect Russian action, given historical patterns. These patterns and the processes driving them have likely shifted or are not the same

---

[30] On this relationship, see Oneal and Russet, 2005. These authors specifically note that the "causes of militarized disputes and wars do not dramatically differ" (p. 299).

(nonstationary). That is, they might not adequately reflect the variables or underlying dynamics that led to more-recent Russian flashpoints, such as the wars with Ukraine and Georgia. We therefore are careful not to over-state the significance of these results.

CHAPTER FIVE

Additional Drivers of Escalation

Recent years have seen potentially escalatory Russian activities that are not captured by the case universe examined in the preceding chapters. The focus of the foregoing chapters on bilateral, dyadic conflict omits potentially escalatory behaviors not reflected in these data sets. In this chapter, we seek to fill this void, exploring potential additional drivers of escalatory activity involving Russia beyond those already examined in earlier chapters of this report.

To do so, we employed a multistage approach. First, we explored instances—both historical and contemporary—of Russian activity outside the previously discussed universe of cases of interstate conflict (or interventions in civil conflicts) to identify categories of activity for examination. Second, we narrowed the scope of our analysis to the subset of these additional drivers that we deemed most likely to engender escalatory activity that could produce a flashpoint, or at least bring the situation *closer* to a flashpoint.[1] Third, we explored several cases of these drivers producing escalatory dynamics in recent years, tracing the processes through which they did so, highlighting variables associated with these drivers that push the situation closer to a flashpoint, and those variables that do not appear to correlate with escalatory dynamics. Finally, we identified potential scenarios involving Russia and the additional drivers of interest that could produce a flashpoint. The sources we consulted to inform this effort included RAND Corporation and other policy or think tank reports, Russian military doctrine, scholarly journal publications, media reports, and conversations with subject-matter experts.

---

[1] This chapter treats flashpoints in line with the definition used throughout the remainder of this report; that is, a *flashpoint* is an instance of a transition from "militarized dispute" to "militarized conflict," involving recorded battle deaths.

Additional Russian Activities That May Drive Escalation

Russia has repeatedly employed nonstate proxies, propaganda, and other types of nonovert action to undermine European security, and scholars have noted that there is good reason to expect that Russia will continue to use these tactics in the future in ways that will threaten core U.S. interests in Europe and around the world.[2] Historically, Russian tactics have often been misleading, clandestine, or covert; Russian actors have regularly sought either misattribution or deniability. Specifically, Russian military plans typically involve *maskirovka* (camouflage and deception), *dezinformatsiya* (disinformation), and/or *refleksivnoe upravlenie* (reflexive control).[3]

Historically, this set of Russian activities takes place at the intersection of unconventional warfare, economic warfare, and involvement in proxy conflicts. Many of the unconventional behaviors that Russia uses to try to gain influence are routine, diffuse, and long-term; others are driven by specific, short-term objectives. Russian tactics can be nonviolent, such as propaganda and disinformation, or they can involve the threat of violence or outright violence.[4]

Using the available literature, we considered the following initial list of possible Russian activities that might drive escalation but are not captured in the previous chapters of this report:

---

[2] Linda Robinson, Todd C. Helmus, Raphael S. Cohen, Alireza Nader, Andrew Radin, Madeline Magnuson, and Katya Migacheva, *Modern Political Warfare: Current Practices and Possible Responses*, Santa Monica, Calif.: RAND Corporation, RR-1772-A, 2018, p. 41.

[3] Ben Connable, Stephanie Young, Stephanie Pezard, Andrew Radin, Raphael S. Cohen, Katya Migacheva, and James Sladden, *Russia's Hostile Measures: Combating Russian Gray Zone Aggression Against NATO in the Contact, Blunt, and Surge Layers of Competition*, Santa Monica, Calif.: RAND Corporation, RR-2539-A, 2020, p. 25.

[4] Stacie L. Pettyjohn and Becca Wasser, *Competing in the Gray Zone: Russian Tactics and Western Responses*, Santa Monica, Calif.: RAND Corporation, RR-2791-A, 2019, p. 21.

- OIE, a term that encompasses a variety of information, cyber, and electromagnetic attacks[5]
- deploying PMCs either covertly or overtly to achieve military, political, or economic objectives[6]
- election meddling
- assassination or targeted killing.

Ultimately, we deemed election meddling and assassination or targeted killing less likely than either OIE or the deployment of PMCs to lead to a flashpoint.[7] The remainder of this chapter therefore explores OIE and

---

[5] We have deliberately chosen OIE rather than *information operations*, which is a subset of OIE typically more focused on *influence* (or nonkinetic or nonlethal effects).

[6] The literature on private military and security contractors is filled with slight differences in the terminology used to refer to these contractors. This reflects the fact that there are often gray areas between different types of contractors; one contracting firm providing mainly logistical support services in one theater of operations might expand its services to include armed security in another theater of operations, muddling the distinction between armed security contractors and other military contractors. In general, a distinction is made between armed private security contractors and the larger category of PMCs (e.g., logistical, base operations support, reconstruction, and security contractors) of which they are a part. This report will focus solely on Russian PMCs—who tend to operate in armed security and paramilitary roles—the generally used terms are *private military company* for firms and *private military contractor* for individual personnel. It should be noted that this use of the terminology is not always consistent in the literature on these topics. Critically, the PMC abbreviation is often used interchangeably to refer to both companies and individual contractors performing military functions for hire; we maintain consistency with this practice here and use the PMC abbreviation interchangeably to denote both individual and corporate entities in this workforce. See Sarah K. Cotton, Ulrich Petersohn, Molly Dunigan, Q. Burkhart, Megan Zander-Cotugno, Edward O'Connell, and Michael Webber, *Hired Guns: Views About Armed Contractors in Operation Iraqi Freedom*, Santa Monica, Calif.: RAND Corporation, MG-987-SRF, 2010.

[7] Although we do not focus separately and directly on election meddling in this chapter because of its relatively low likelihood of leading to a flashpoint compared with other drivers, PMC activity appears to be related to election meddling to some extent. Wagner Group PMCs might have been involved in meddling or at least intended to meddle in the August 2020 elections in Belarus. Additionally, Russian operatives with ties to Russian business mogul Yevgeny Prigozhin—widely considered to be the head of Wagner Group—have attempted to cultivate influence in Libya and sought to promote Seif al-Islam al-Gaddafi, son of the country's former dictator, as a candidate in prospective

the deployment of PMCs as two potential additional drivers of escalatory dynamics that could produce flashpoints with Russia.

Both OIE and the utilization of PMCs pose challenges for attribution of hostile activities. Attribution does not simply mean identifying the actor responsible for an attack; it could mean directly connecting a proxy to a state and then proving direction of the proxy by the state. This can be exceptionally difficult, given the relative ease of maintaining plausible deniability in both contexts. Attribution is critical because it enables decisionmakers to consider appropriate responses to Russian behavior. Furthermore, without attribution or explicit acknowledgement, assertive behavior could lead to inadvertent escalation through misattribution or misunderstanding. The risk might be even higher when horizontal escalation is used because that tactic further muddies the waters in terms of divining intentions.[8]

Russian Operations in the Information Environment

In recent years, the U.S. Department of Defense (DoD) has adopted the phrase *operations in the information environment* as an umbrella term that refers to a variety of statutorily and doctrinally distinct activities in the information realm. The DoD *Strategy for Operations in the Information Environment* describes OIE as activities occurring in the information environment,[9] which the DoD *Dictionary of Military and Associated Terms*

---

presidential elections. One of the operatives had also been on a team of Russians who had overtly and unsuccessfully meddled in Madagascar's 2018 elections on behalf of former president Hery Rakotoarimanana. See "Belarus Arrests Dozens of Russian Mercenaries: State Media," *Al Jazeera*, July 29, 2020; Andrew Higgins and Declan Walsh, "How Two Russians Got Caught Up in Libya's War, Now an Action Movie," *New York Times*, June 18, 2020; and Michael Schwirtz and Gaelle Borgia, "How Russia Meddles Abroad for Profit: Cash, Trolls and a Cult Leader," *New York Times*, November 11, 2019.

[8] For more on escalation, see Forrest E. Morgan, Karl P. Mueller, Evan S. Medeiros, Kevin L. Pollpeter, and Roger Cliff, *Dangerous Thresholds: Managing Escalation in the 21st Century*, Santa Monica, Calif.: RAND Corporation, MG-614-AF, 2008.

[9] DoD, *Department of Defense: Strategy for Operations in the Information Environment*, Washington, D.C., June 2016.

defines as "the aggregate of individuals, organizations, and systems that collect, process, disseminate, or act on information."[10]

OIE began to garner widespread media attention following the Russian interference in the 2016 U.S. presidential election. Such influence operations, although certainly problematic, have not come close to producing a flashpoint-like situation. The OIE we analyze here—cyberattacks and electromagnetic attacks (EAs)—by contrast, are more-plausible sparks for flashpoints. However, even these forms of OIE have yet to lead to a flashpoint in practice. The lack of a case of a Russian OIE leading to a flashpoint in any historical cases thus far is in itself an interesting, if preliminary, finding. We discuss this in further detail later in this chapter and explore a set of descriptive variables that could produce escalatory dynamics leading to a flashpoint in the future.

As defined by the DoD dictionary, *cyberattacks* are "actions taken in cyberspace that create noticeable denial effects (i.e., degradation, disruption, or destruction) in cyberspace or manipulation that leads to denial that appears in a physical domain, and is considered a form of fires."[11] Of the two behaviors we examine in this section, cyberattacks are the most revolutionary. There are at least three reasons for this. First, they provide an attacker with unmatched speed and global reach.[12] Second, although cyberattacks

---

[10] DoD, *DoD Dictionary of Military and Associated Terms*, Washington, D.C., January 2021, p. 104. The *Joint Concept for Operating in the Information Environment* (JCOIE) expands on this definition, noting that the information environment revolves around "three interrelated but distinct dimensions" (DoD, *Joint Concept for Operating in the Information Environment (JCOIE)*, Washington, D.C., July 25, 2018, p. 2). These dimensions are the

> physical dimension, where information overlaps with the physical world; the information dimension, where information is collected, processed, stored, disseminated, displayed, and protected, including both the content and the flow of information between nodes; and the cognitive dimension, where human decisionmaking takes place based upon how information is perceived (Catherine A. Theohary, *Defense Primer: Information Operations*, Washington, D.C.: Congressional Research Service, IF10771, December 15, 2020, p. 1).

[11] DoD, 2021, p. 55. We use the term *cyberattack* because it is a commonly accepted simplified variation of "cyberspace attack." *Fires* refers to the use of weapon systems or other actions to create specific lethal or nonlethal effects on a target.

[12] Timothy L. Thomas, "Deterring Information Warfare: A New Strategic Challenge," *Parameters*, Vol. 26, No. 4, Winter 1996–97.

can be precisely targeted, they often cause considerable and unpredictable collateral damage.[13] Third, tools employed in cyberspace present a low barrier to entry and offer relative ease to mask attribution (unlike, for example, intercontinental ballistic missiles—which possess similar range and stand-off capability). This makes cyberattacks a desirable capability for both states and nonstate actors.[14]

Of the countless cyberattacks over the past two decades, several are notable for their innovative impact on the character of warfare, but none has captured the public's imagination quite like Stuxnet. Stuxnet, a joint endeavor by Israel and the United States against Iran in the mid-2000s, is generally seen as the first cyberattack on critical industrial infrastructure.[15] Many believe it caused Iran to start taking the cyber domain more seriously, ultimately leading it to become one of the world's most significant cyber powers.[16] Notably, the attack was narrowly targeted at Iran's nuclear program and did not appear to harm people (or be intended to achieve such an effect).[17] Another milestone in this domain were the cyberattacks on Ukraine's power grid in 2015 and 2016, the first to cause blackouts. It is now clear that Russia-based hackers were behind these attacks, though their

---

[13] For example, the White House called NotPetya "the most destructive and costly cyberattack in history," after it went global instead of just affecting its intended target in Ukraine (Anton Troianovski and Ellen Nakashima, "How Russia's Military Intelligence Agency Became the Covert Muscle in Putin's Duels with the West," *Washington Post*, December 28, 2018).

[14] Quentin E. Hodgson, Logan Ma, Krystyna Marcinek, and Karen Schwindt, *Fighting Shadows in the Dark: Understanding and Countering Coercion in Cyberspace*, Santa Monica, Calif.: RAND Corporation, RR-2961-OSD, 2019, p. 2.

[15] David E. Sanger, *Confront and Conceal: Obama's Secret Wars and Surprising Use of American Power*, New York: Crown Publishing Group, Crown Publishers, 2012, p. 188.

[16] Kate Fazzini, "The Saudi Oil Attacks Could Be a Precursor to Widespread Cyberwarfare—with Collateral Damage for Companies in the Region," *CNBC*, September 21, 2019.

[17] Such intentionality should not be seen as definitively determinative of escalation potential; inadvertent escalation is possible (P. W. Singer, "Stuxnet and Its Hidden Lessons on the Ethics of Cyberweapons," *Case Western Reserve Journal of International Law*, Vol. 47, No. 1, 2015).

objectives remain unclear.[18] Although there was no indication of an intent to cause harm to people, both attacks occurred in the middle of winter and could have indirectly led to such an outcome.

The 2017 attack against a Saudi oil refinery demonstrated even greater escalatory potential.[19] Although it appears that no cyberattack *in itself* to date can be traced to a combat death or a flashpoint between states, it might very well be only a matter of time until such an attack does precipitate a flashpoint, as cyber practitioners continue to learn from each other and refine their own tools and methods.

The DoD dictionary defines an *EA* as "the use of electromagnetic energy, directed energy, or antiradiation weapons to attack personnel, facilities, or equipment with the intent of degrading, neutralizing, or destroying enemy combat capability and is considered a form of fires."[20] Much like cyberattacks, EAs, when executed properly, can make attribution challenging. Attribution is complicated because an EA tends to be unannounced and covert, so the victim might not even be aware of being under attack while experiencing the attack's effects.

Russian military electronic warfare (EW) has experienced a renaissance in the past decade after atrophying significantly in the 1990s. Combat opportunities in Syria and Ukraine, among other locations, have afforded critical opportunities to exercise, test, and refine recent investments in equipment and structural reforms related to EW. The Russian military now possesses a broad array of platforms, both air and ground, that covers the variety of EW missions from electromagnetic protection to EA. In Syria, it has used both electronic countermeasure (ECM) pods on its aircraft and ground-based jammers to thwart adversary radar in support of air operations. By contrast, in Ukraine, it used ground-based platforms to jam, intercept, and monitor communications.[21]

---

[18] Ben Buchanan, "Five Myths About Cyberwar," *Washington Post*, February 20, 2020.

[19] Andy Greenberg, "The Highly Dangerous 'Triton' Hackers Have Probed the U.S. Grid," *Wired*, June 14, 2019.

[20] DoD, 2021, p. 69.

[21] Roger N. McDermott, *Russia's Electronic Warfare Capabilities to 2025: Challenging NATO in the Electromagnetic Spectrum*, Tallinn, Estonia: International Center for Defence and Security, September 2017.

Although mostly targeted at adversary military targets, EAs can affect civilians, especially communication and navigation of commercial ships, aircraft, and emergency services personnel.[22] It is also important to note that EW support can be used to facilitate kinetic targeting, as seen in Ukraine, with the use of direction-finding equipment.[23]

We examine two particular OIE cases in more detail here. These were selected because they are well documented, came closest to causing a flashpoint, and did not occur in the context of an ongoing flashpoint. This last criterion is important because OIE can often play a supporting role in interstate conflict, as was the case in Russia's wars with Ukraine and Georgia.[24] Here, we are interested in assessing the independent escalatory potential of OIE; thus, we avoided cases where these actions played a supporting role.

Operations in the Information Environment Case Vignettes

Triton (Cyberattack)

In August 2017, Saudi Aramco, the largest oil company in the world, suffered a cyberattack intended to cause the safety shutoffs at the Petro Rabigh refinery to fail and, in turn, cause an explosion. Had the attack succeeded, it would have been the first of its kind; it would have unquestionably inflicted human casualties.[25] Instead, thanks to a flaw in the code used for the attack, the crisis was averted.[26]

This was not the first cyberattack experienced by Saudi Arabia, Aramco, or the Petro Rabigh refinery. In 2012, a hacking group known as Xeno-

---

[22] Hodgson et al., 2019, p. 26; Norwegian Intelligence Service, *Focus 2019: The Norwegian Intelligence Service's Assessment of Current Security Challenges*, Oslo, January 21, 2019; and Nicole Perlroth and Clifford Krauss, "A Cyberattack in Saudi Arabia Had a Deadly Goal. Experts Fear Another Try," *New York Times*, March 15, 2018.

[23] McDermott, 2017, p. 25.

[24] See McDermott, 2017.

[25] David E. Sanger, "Hack of Saudi Petrochemical Plant Was Coordinated From Russian Institute," *New York Times*, October 23, 2018.

[26] Zahraa Alkhalisi, "Security Experts: Iran-Backed Hackers Targeting U.S. and Saudi Arabia," *CNN Business*, September 21, 2017.

time, or the Triton Actor, conducted a cyberattack with malware known as "Shamoon." Shamoon destroyed 75 percent of the massive refinery's tens of thousands of computers. Fortunately, those computers were for general business and were segregated from the refinery's control system computers, so both risk and impact were relatively limited.[27]

In the 2017 attack, however, a new and improved Shamoon managed to punch through an ineffective firewall, gain access to corporate computers, and subsequently migrate to operational technology—the refinery's control system. Neither the control system computers nor the plant floor signaled alarm to the engineers in the control room.[28] More specifically, Shamoon went after the Triconex controllers, manufactured by Schneider Electric, "which keep equipment operating safely by performing such tasks as regulating voltage, pressure and temperatures. Those controllers are used in about 18,000 plants around the world, including nuclear and water treatment facilities, oil and gas refineries, and chemical plants."[29] Ultimately, these devices serve a role akin to home circuit breakers: averting catastrophe in advance of human intervention. The goal of the attack was to cause the controllers to fail, resulting in an explosion in one of the largest facilities of its kind in the world.[30] The cybersecurity firm Dragos has called the Triton actor "easily the most dangerous threat activity publicly known."[31] It is also worth noting that the firm has evidence that the same group "has probed the networks of at least 20 different U.S. electric system targets," creating vulnerabilities to exploit in a future attack.[32]

Initially, Iran was suspected as the likely source of the attack because of its highly adversarial relationship with Saudi Arabia. But over a year later, the cybersecurity firm FireEye was able to trace the code back to the Russian Central Scientific Research Institute of Chemistry and Mechanics, a

---

[27] Hodgson et al., 2019, p. 25.

[28] Blake Sobczak, "The Inside Story of the World's Most Dangerous Malware," *E&E News*, March 7, 2019.

[29] Perlroth, 2018.

[30] Sobczak, 2019.

[31] Greenberg, 2019.

[32] Greenberg, 2019.

government-owned institution in Moscow. FireEye further asserted that the Russian government had direct involvement in the attack.[33] Although attribution was ultimately determined in this case (albeit months after the attack), the objective of the attack remains disturbingly unclear. The Russian attackers could have merely been experimenting with new capabilities, or they could have intended a targeted attack reflecting Russia's competition with Saudi Arabia over global energy pricing and markets. However, these are speculative hypotheses, and it is equally plausible to argue that the attack lacked a policy-driven logic, given recent close collaboration between Riyadh and Moscow. As FireEye's intelligence director put it, "Sometimes [cyberattacks make] no geopolitical sense."[34]

Trident Juncture 18 (EA)

Trident Juncture 18 (TRJE18) was the largest NATO exercise since the end of the Cold War. It featured "40,000 troops from all 29 NATO members, along with 70 ships, 150 aircraft, and 10,000 ground vehicles."[35] It took place across Northern Europe in October and November 2018, with the preponderance of ground activities in Norway; there were also maritime activities in the Baltic and North Seas. Along with NATO members, Sweden and Finland also participated.[36]

In response, Russia jammed global positioning system (GPS) and communication signals in the high north while TRJE18 was ongoing from a base on the Kola peninsula.[37] Figure 5.1 shows the scope of jamming across Norwegian and Finnish territory and a bar graph indicating that jamming

---

[33] Dustin Volz, "Researchers Link Cyberattack on Saudi Petrochemical Plant to Russia," *Wall Street Journal*, October 23, 2018.

[34] Sanger, 2018.

[35] Paul McLeary, "Russians Tried to Jam NATO Exercise; Swedes Say They've Seen This Before," *Breaking Defense*, November 20, 2018.

[36] McLeary, 2018.

[37] Brooks Tigner, "Electronic Jamming Between Russia and NATO Is Par for the Course in the Future, but It Has Its Risky Limits," Atlantic Council, November 15, 2018. This was not the first time that such an incident had occurred; similar activity took place in advance of and during the Russian exercise Zapad in September 2017, though with less impact (Elisabeth Braw, "The GPS Wars Are Here," *Foreign Policy*, December 17, 2018).

FIGURE 5.1

Scope and Timeline of Russian Jamming During Trident Juncture 18

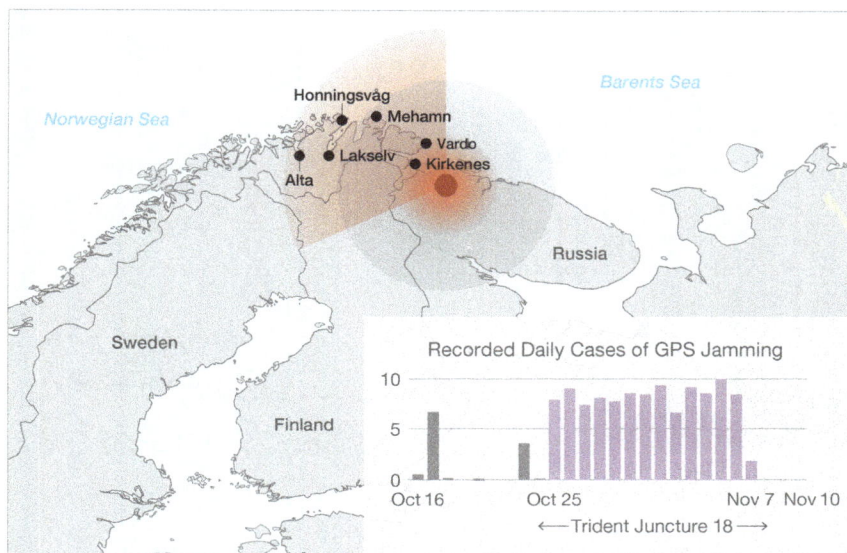

SOURCE: Kjell Persen, "Intelligence Service: This Is How Norway Was Jammed During the NATO Exercise," *TV2 Norway*, February 11, 2019.

activity correlated directly with the execution of TRJE18. In a public brief-ing three months after the exercise, Morten Haga Lunde, the director of Norway's military intelligence service, explained that the Russian military moved a jamming system to an airbase only 16 km from the Russia-Norway border in mid-October 2018. Immediately thereafter, the Norwegians began experiencing sporadic jamming. However, the Russian military then moved the equipment to higher terrain, and the following day, in conjunction with the start of TRJE18, jamming dramatically increased in scope and strength. The jamming concluded on the same day as the exercise, at which point the Russian system returned to its garrison (see Figure 5.2).[38]

The attacks caused both Norwegian and Finnish civilian aircraft to lose GPS signal, requiring the use of alternative navigation systems and, in some

---

[38] Persen, 2019.

FIGURE 5.2

Trident Juncture 18: Russian Jamming Timeline and Locations

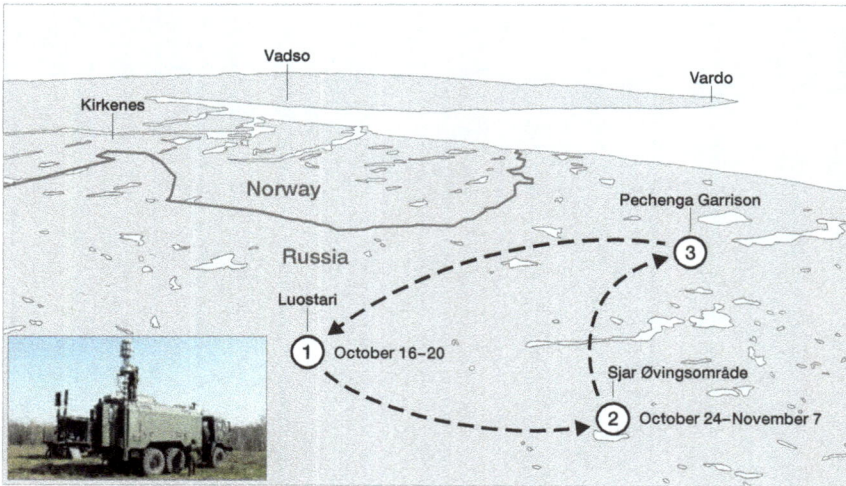

SOURCE: Persen, 2019.

cases, necessitating rerouting.[39] Norway had no doubt about the source and was not shy about pointing the finger at Russia. Defense Minister Frank Bakke-Jensen told Norwegian Radio that "Russia doesn't respect civilian aviation in the North."[40] Finland took it a step further and summoned the Russian ambassador to explain why commercial airliners were reporting instances of jamming.[41]

According to Timothy Thomas, this EA presents a clear example of Russia executing its "disorganization concept" to deliberately disrupt command and control.[42] Russia in recent years has increasingly used EW in a variety of ways, such as jamming cell phone networks, hacking service members'

---

[39] Persen, 2019.

[40] Braw, 2018.

[41] Gerard O'Dwyer, "Finland to Bolster Navy's Surface Fleet with New Ships, More Missiles," *Defense News*, January 13, 2020.

[42] Timothy L. Thomas, *Russian Military Thought: Concepts and Elements*, McLean Va.: MITRE Corporation, MP190451V1, August 2019.

personal devices, and jamming U.S. forces in Syria.[43] The EA on TRJE18 differs from these examples of previous cases of Russian EW because of the apparent intention to create effects on both the military and civilian sectors of the adversary—with potentially catastrophic consequences.[44]

Analysis: Operations in the Information Environment as a Potential Driver of Escalatory Dynamics

Table 5.1 summarizes the two cases according to descriptive variables using the information from the vignettes we have presented. Although neither produced a flashpoint, these two OIE (among the hundreds, perhaps even

TABLE 5.1

Operations in the Information Environment Cases Summary

Attribute	Triton	Trident Juncture 18
Background		
Form of OIE	Cyber	EA
Location	Saudi Arabia	Norway/Finland
Year	2017	2018
Target	Oil refinery	Communication and navigation (military and civilian)
Attribution		
Attribution clear at the time of the attack	No	Yes
Government directed	Likely	Yes
Effects		
Physical damage (intended)	Likely	Unlikely
Casualties (intended)	Likely	Unlikely
Physical damage (achieved)	No	No
Casualties (achieved)	No	No

---

[43] Tigner, 2018.

[44] Tigner, 2018.

thousands, that Russia has executed in recent years) appear to have come closest to provoking an interstate conflict.

It is difficult to posit precisely the circumstances under which OIE could lead to a flashpoint, given the lack of historical precedent. That said, a detailed examination of the attributes of these two cases offers some insight into what characteristics of these operations produce escalatory dynamics.

First, a victim is less likely to react to an attacker in a way that could produce an escalatory dynamic if the attack cannot be attributed in a timely fashion. The Triton attack was originally misattributed to Iran; Russia was finally outed over a year after the fact. At that point, a tit-for-tat escalation was unlikely even if Saudi Arabia did covertly retaliate. Without prompt and accurate attribution, interstate frictions are far less likely. Additionally, in the cyber context, attribution to a state actor is also significant. Attacks executed by nonstate hacker groups are less likely to produce retaliation against a state.

Effects are probably the biggest factor in whether OIE lead to a flashpoint. Physical damage raises the stakes for any nation, especially if intentionally inflicted. Although there was no such damage associated with the two OIE cases that we have examined here, escalation would have been more likely if physical damage had occurred in either case. Kinetic or lethal effects in any domain often are the ultimate red line in interactions between states.

To summarize the lessons from these Russian OIE cases, prompt, accurate attribution and the severity of the effects of the operation seem to increase the risk that the state on the receiving end of a Russian attack will retaliate. Retaliation, in turn, risks driving an escalatory spiral that could eventually lead to interstate conflict. An attributed attack that inflicted physical damage is probably much more likely to escalate to a flashpoint (assuming the victim retaliates) than an attack that does not cause damage or is not attributed in a timely fashion.

Activities of Russian Private Military Contractors

The extent to which the Russian state and state-affiliated entities are using PMCs has exploded in the past decade, and these actors are increasingly thought to constitute a critical component of the Russian military ground operations. PMCs are enabling covert, plausibly deniable Russian activity in proxy conflicts in Europe, Africa, Latin America, and—most notably—in Syria, enabling expansion of Russia's global strategic footprint. Perhaps the most prominent Russian PMC, the so-called Wagner Group, is or has recently been active in Syria, Sudan, the Central African Republic (CAR), Madagascar, Libya, and Mozambique.[45] By 2017, the Wagner Group might have employed as many as 5,000 contractors to operate just in Syria.[46] Though the most widely covered in the press, the Wagner Group is just one of several known Russian PMCs that have formerly operated and/or continue to operate abroad.[47]

Media and scholarly reports provide a mixture of hard evidence and anecdotal speculation in their detailing of how contractors working under the auspices of one or more of these companies have reportedly operated—some as far back as the late 1990s—in Iraq, Afghanistan, Ukraine, Syria, Libya, Chechnya, Tajikistan, Yemen, and Burundi. This is in addition to the Wagner Group operations.[48]

---

[45] Ryan Browne, "Top U.S. General Warns Russia Using Mercenaries to Access Africa's Natural Resources," *CNN*, February 7, 2019; subject-matter expert presentation at Columbia University Workshop on U.S.-Russia Relations, New York, June 26, 2019; Jane Flanagan, "Mozambique Calls on Russian Firepower," *The Times* (London), October 2, 2019; Tim Lister, Sebastian Shukla, and Clarissa Ward, "CNN Special Report: Putin's Private Army," *CNN*, August 2019; and Kimberly Marten, "Russia's Use of Semi-State Security Forces: The Case of the Wagner Group," *Post-Soviet Affairs*, Vol. 35, No. 3, 2019.

[46] Sarah Fainberg, "Russian Spetsnaz: Contractors and Volunteers in the Syrian Conflict," *Russie.Nei.Visions*, No. 105, French Institute of International Relations, December 2017; and Neil Hauer, "Putin Has a New Secret Weapon in Syria: Chechens," *Foreign Policy*, May 4, 2017. Other reports were more precise, detailing slightly lower figures of 4,840 Wagner contractors, though it is unclear whether all of those were deployed to Syria at that time (Sergey Sukhankin, "War, Business and 'Hybrid' Warfare: The Case of the Wagner Private Military Company [Part Two]," *Eurasia Daily Monitor*, Vol. 15, No. 61, April 23, 2018a).

[47] Other Russian PMCs that have appeared in media and social media reporting since 2001 are Moran Security Group, RSB Group, Slavonic Corps, E.N.O.T. Corps, MAR, and Centre R.

[48] Notably, Wagner Group operations are arguably more paramilitary and directly linked to the Russian state than are the activities of at least some of these other firms. The data on this issue are incomplete and difficult to access, but there is some evidence to indicate that Wagner has a privileged, formal relationship with the Russian state that

Despite this diversity of firms operating in the Russian market for force, the contractors working for these firms are almost uniformly of Russian, Serbian, Ukrainian, or Moldovan descent.[49] Like their Western PMC employee counterparts, most are either former servicemen or former officers of Russia's military intelligence agency (commonly referred to as the GRU) and the Federal Security Service, a domestic intelligence and law-enforcement agency.[50] Unlike most Western PMCs, Russian contractors have reportedly been engaged in combat missions (both by the Russian state and by other actors), operating more like paramilitary forces than private security providers.[51]

Interestingly, PMCs—and utilization of their services—are technically illegal in Russia, despite extensive evidence of close Russian state support and client relationships with particular firms.[52] Prominent figures, including Putin, have argued for legalization, and a bill to legalize PMCs was proposed in 2018 to Russia's Duma but did not pass.[53]

Private Military Contractor Case Vignettes

To elicit a better understanding of how, where, and when Russian private military actors might pose a greater risk of escalating interstate tensions or creating a flashpoint, we explore four comparative case vignettes. Although hindered from making sweeping claims regarding causality by the small-n nature of the comparison and the data limitations inherent in any research

---

other such firms might not have enjoyed historically. See Browne, 2019; Flanagan, 2019; Lister, Shukla, and Ward, 2019; and Marten, 2019.

[49] Subject-matter expert presentation at Columbia University Workshop on U.S.-Russia Relations, New York, June 26, 2019.

[50] Andrew Linder, *Russian Private Military Companies in Syria and Beyond*, Washington, D.C.: Center for Strategic and International Studies, undated.

[51] Emmanuel Dreyfus, *Private Military Companies in Russia: Not So Quiet on the Eastern Front*, Paris: Insitut de Recherche Stratégique de L'École Militaire, October 12, 2018.

[52] Marten, 2019.

[53] Sergey Sukhankin, "'Continuing War by Other Means': The Case of Wagner, Russia's Premier Private Military Company in the Middle East," Jamestown Foundation, July 13, 2018b.

on this topic, we identify a set of variables that appear to correlate with an increased likelihood of near-flashpoint situations involving the Russian state. The first two cases produced combat deaths, though neither led to an overt interstate conflict.

Deir Ezzor

Russian PMCs have been a component of the Syrian battlespace for years.[54] They came closest to generating interstate conflict on February 8, 2018. In an attempt to seize a Conoco gas plant east of the Euphrates River in Syria's Deir Ezzor province, a battalion-sized formation of Wagner personnel and pro-regime Syrian militiamen (approximately 500 fighters) attacked a position held by U.S. forces and their partners, the Syrian Democratic Forces (SDF), with artillery, tanks, and mortars. After contacting Russian military counterparts via the U.S.-Russian deconfliction hotline and receiving assurances that Russian state forces were not involved, U.S. forces responded decisively. Over the course of four hours, a combination of fighter jets, bombers, tactical unmanned aerial vehicles (UAVs), and attack helicopters pummeled the attacking force. By the time the PMCs and their Syrian partners withdrew back across the Euphrates, between 200 and 300 people had reportedly been killed. For their part, no U.S. service members were injured, and only one SDF fighter was wounded.[55]

Follow-on strikes and sporadic clashes between the Syrian regime and U.S.-aligned forces continued in the area for another week. However, fears that the February 8 clash would escalate to a wider conflagration between Russia and the regime on one side and the United States on the other ulti-

---

[54] For example, 267 personnel of the Slavonic Corps deployed to Syria in 2013, initially to guard Assad regime-held oil infrastructure. However, they were quickly involved in firefights with anti-regime elements, suffered heavy casualties, and withdrew from the country. On returning to Russia, two of the Slavonic Corps' commanders were imprisoned; nevertheless, some of those same fighters would eventually return to Syria as part of a different PMC: the Wagner Group (Kimberly Marten, "The Puzzle of Russian Behavior in Deir al-Zour," *War on the Rocks*, July 5, 2018; and Candace Rondeaux, *Decoding the Wagner Group: Analyzing the Role of Private Military Security Contractors in Russian Proxy Warfare*, Washington, D.C.: New America, November 7, 2019, p. 45).

[55] Thomas Gibbons-Neff, "How a 4-Hour Battle Between Russian Mercenaries and U.S. Commandos Unfolded in Syria," *New York Times*, May 24, 2018.

mately proved unfounded. Regardless, the ambiguity surrounding the Russian government's prior knowledge of and involvement in Wagner's actions—some reports suggested the PMCs were operating on behalf of the regime to retake hydrocarbon-related assets—reflect the risk of Russian PMCs operating in a highly dynamic battlespace.

Tripoli

Russia has long been involved in Libya's civil war, providing support to eastern-based strongman Khalifa Hifter, his self-styled Libyan Arab Armed Forces (LAAF), and the eastern-based Libyan House of Representatives.[56] However, Russian involvement in the Libyan conflict increased significantly in fall 2019 when an estimated 1,200 Wagner personnel deployed to support Hifter's ongoing assault on Tripoli, Libya's capital and seat of the internationally recognized Government of National Accord.[57]

Initially planning to seize the capital rapidly, Hifter's offensive—launched April 4, 2019—quickly bogged down in Tripoli's southern outskirts, and LAAF-affiliated forces failed to make further substantive gains. However, beginning in September 2019, the increased involvement of Wagner personnel, directing and participating in LAAF operations in and around Tripoli, led to some of the LAAF's largest territorial gains in almost six months. In addition to serving as snipers and forward operators coordinating indirect fires, Wagner personnel reportedly embedded in LAAF operation centers and might even have influenced the LAAF's force structure in the Tripoli campaign.[58] Moreover, the high level of capabilities and operational sophistication Wagner brought to the battlespace reportedly had a significant psychological impact on Tripoli's defenders, who became increasingly beleaguered.[59] Hifter-aligned forces' momentum was only halted and reversed

---

[56] Anas El Gomati, "Russia's Role in the Libyan Civil War Gives It Leverage over Europe," *Foreign Policy*, January 18, 2020.

[57] Michelle Nichols, "Up to 1,200 Deployed in Libya by Russian Military Group: U.N. Report," Reuters, May 6, 2020.

[58] Interview with U.S. think tank researcher, Washington, D.C., May 5, 2020.

[59] Nathan Vest and Colin P. Clarke, "Is the Conflict in Libya a Preview of the Future of Warfare?" *Defense One*, June 2, 2020; and Frederic Wehrey, "With the Help of Russian Fighters, Libya's Haftar Could Take Tripoli," *Foreign Policy*, December 5, 2019.

by Turkey's military intervention on behalf of the Government of National Accord in January 2020, in part a direct response to Wagner's involvement in the battle for Tripoli. By mid-2020, Hifter's offensive had collapsed.[60]

Wagner's activities in Tripoli also engendered increased U.S. attention to the Libyan conflict. Bipartisan bills were introduced in both the House and Senate, which, inter alia, called for highlighting foreign meddling in Libya and establishing a strategy to counter Russian operations in the country. To date, the U.S. military has not directly engaged Wagner in Libya. (There are reports that a Turkish drone strike killed 35 Wagner mercenaries in September 2019.)[61] However, the United States accused Wagner of downing a U.S. MQ-9 UAV over Tripoli in late November 2019, suggesting that the group's personnel in Tripoli might have possessed more-sophisticated equipment than their counterparts in Deir Ezzor.[62]

Central African Republic

Commercial and security ties between Russia and the CAR—embroiled by civil war since 2012—began strengthening in October 2017 when President Putin met with his CAR counterpart, Faustin-Archange Touadéra, in Sochi, Russia, to discuss aid to the beleaguered government in Bangui. Soon thereafter, two Russian-owned firms with ties to Prigozhin—the mining company Lobaye Invest and security firm Sewa Security Services—registered in Bangui.[63] In December 2017, the UN granted Russia an exemption to provide arms and training to the Central African Armed Forces (FACA) and Presidential Guard, which have been under a UN arms embargo since 2013. To support CAR state forces, Russia deployed 175 trainers, five of whom

---

[60] International Crisis Group, *Turkey Wades into Libya's Troubled Waters*, Brussels, April 30, 2020.

[61] Andrew McGregor, "Falling off the Fence: Russian Mercenaries Join the Battle for Tripoli," *Eurasia Daily Monitor*, Vol. 16, No. 138, October 8, 2019.

[62] Shawn Snow, "AFRICOM Demands Return of U.S. Drone Shot Down by Russian Air Defenses over Libya," *Military Times*, December 10, 2019.

[63] Lyubov Glazunova, "Russia Is Washing Blood Off African Diamonds," RIDDLE, June 9, 2020.

were Russian state military service members and 170 of whom were PMCs and suspected of being Wagner personnel.[64]

The Russian security actors' primary and official role in the CAR was training and supporting the FACA, enabling it to reassert governmental control throughout the country. Russian-led training activities have taken place both inside the CAR (the Lobaye Prefecture outside Bangui) and outside the country (in Sudan).[65] Russian PMCs have also guarded infrastructure (including hospitals and diamond mines), served as President Touadéra's personal security detail, and accompanied FACA elements in conducting mass arrests, which have resulted in allegations of torture and other human rights abuses.[66]

Though Moscow is operating under the auspices of supporting the Bangui government and resolving the long-running civil war, observers of Russia's activities in the CAR have raised concerns that Moscow is providing security assistance as a mechanism of pursuing parallel commercial and political interests.[67] In particular, Moscow is likely leveraging its growing influence in the CAR to obtain more access to the country's lucrative diamond industry, which Wagner's ground presence facilitates. In June 2018, the Prigozhin-linked Lobaye Invest mining company gained exploration rights to mines previously under rebel control, reflecting a possible incentive-

---

[64] Romain Esmenjaud, Mélanie De Groof, Paul-Simon Handy, Ilyas Oussedik, and Enrica Picco, *Midterm Report of the Panel of Experts on the Central African Republic Extended Pursuant to Security Council Resolution 2399*, New York: United Nations Security Council, July 23, 2018, p. 7. The 2019 UN panel of experts' report stated that the number of Russian trainers had risen to 235; however, it does not clarify how many of those trainers are Russian state military personnel (Romain Esmenjaud, Mélanie De Groof, Ilyas Oussedik, Anna Osborne, and Émile Rwagasana, *Final Report of the Panel of Experts on the Central African Republic Extended Pursuant to Security Council Resolution 2454*, New York: United Nations Security Council, December 14, 2019, p. 33).

[65] Esmenjaud et al., 2018, p. 7.

[66] Dionne Searcey, "Gems, Warlords, and Mercenaries: Russia's Playbook in Central African Republic," *New York Times*, September 30, 2019.

[67] Mike Eckel, "New Scrutiny for 'Putin's Chef' and Russian Mercenaries in Africa," Radio Free Europe/Radio Liberty, October 1, 2019; Glazunova, 2020; Leslie Minney, Rachel Sullivan, and Rachel Vandenbrink, "Amid the Central African Republic's Search for Peace, Russia Steps In. Is China Next?" United States Institute for Peace, December 19, 2019.

based structure by which companies linked to Wagner's owners gain concessions to diamond mines as Wagner contractors help bring the mines under governmental control.[68] In 2018, President Touadéra appointed a former Russian intelligence officer, Valery Zakharov, as his security adviser, likely yielding to Russia substantial influence over Central African affairs.[69]

Cabo Delgado (Mozambique)

Moscow and Maputo have long-standing relations, dating back to the Cold War when the Soviet Union supported Mozambican independence from Portugal. Although Russian ties with Mozambique diminished after the collapse of the Soviet Union, Russia has sought in recent years to reengage across Africa, and Mozambique is no exception.

In September 2019, 200 Russian security personnel deployed to Mozambique to support the Mozambique Armed Defense Forces (FADM) in combatting a mounting Islamic State of Iraq and Syria (ISIS)–linked insurgency in the country's north. Media reports of the deployment first suggested that Russian military forces had arrived in Mozambique; however, it was later confirmed that the Russian operatives were actually Wagner contractors.[70] The Wagner personnel reportedly deployed with advanced military equipment—relative to other actors in the theater—such as helicopters, UAVs, and armored personnel carriers. However, the contractors seemed tactically ill prepared for operating in Mozambique's harsh bush environment. Within weeks of arriving in theater, several Wagner personnel were

---

[68] Three Russian journalists who were investigating alleged Russian profiteering from so-called blood or conflict diamonds were killed in July 2018 in a mine outside Bangui. See Eckel, 2019; Glazunova, 2020.

[69] Searcey, 2019. Although attending the 2019 Russia-Africa Summit in Sochi, President Touadéra even stated that his government was open to hosting a Russian military base in the CAR. See Andrew Roth, "Central African Republic Considers Hosting Russian Military Base," *The Guardian*, October 25, 2019.

[70] Sergey Sukhankin, "Russian PMCs and Irregulars: Past Battles and New Endeavors," Jamestown Foundation, May 13, 2020d; and Al J. Venter, "A Dirty Little War in Mozambique," *Key Aero*, No. 386, May 2020. Of note, Wagner reportedly beat out at least two other PMCs—one owned by a South African and the other by a Rhodesian—that were vying for the Mozambican contract to support the FADM against northern insurgents (Sukhankin, 2020d).

killed in insurgent ambushes; after multiple failed operations, Wagner and FADM elements reportedly ceased joint bush patrols.[71]

Wagner's effort to combat the ISIS-affiliated insurgency was an abject failure; U.S. Africa Command commander General Stephen Townsend called it "second-rate counterterrorism assistance."[72] Nevertheless, Russian firms still reportedly have plans for the country's lucrative and underexplored natural gas reserves, much of which are located in the northern, insurgency-afflicted regions. Analyst Sergey Sukhankin has framed Moscow's approach in Mozambique as the "Syria Model," by which Moscow provides security assistance in exchange for concessions in Mozambique's energy sector.[73]

Analysis: Private Military Contractors as a Potential Driver of Escalation

Table 5.2 compares the variables associated with the cases of PMC deployment to isolate those associated with greater potential for such a deployment to escalate to a flashpoint. This comparison suggests that two variables in particular, presence of a strong adversary's forces and PMC tactical engagement, might increase the risk of escalation. Unsurprisingly, PMC presence is unlikely to drive escalatory dynamics between Russia and another state unless the PMCs are operating on a tactical level and an adversary's state has uniformed personnel in the area of operation whom the PMCs are threatening. As seen in Deir Ezzor, if Russian PMCs threaten a capable adversary's uniformed forces, that state is likely to defend its personnel and respond with force.

---

[71] Pjotr Sauer, "7 Kremlin-Linked Mercenaries Killed in Mozambique in October—Military Sources," *Moscow Times*, October 31, 2019; and Venter, 2020, p. 77.

[72] Stephen J. Townsend, "Statement of General Stephen J. Townsend, United States Army Commander, United States Africa Command, Before the Senate Armed Services Committee," Washington, D.C., January 30, 2020, p. 16.

[73] Sukhankin, 2020d. In August 2019, Moscow forgave 95 percent of Maputo's debt in exchange for increased Russian investment opportunities in the country, and Maputo granted Russian oil and gas giant Rosneft rights to study geographical data of possible gas deposits ("Russian Military Personnel in Mozambique: Bringing Another African Nation Closer to Moscow," Warsaw Institute, September 30, 2019).

TABLE 5.2

Private Military Contractor–Related Variables Possibly Associated with a Heightened Risk of Escalation

Variable	Deir Ezzor	Tripoli	CAR	Cabo Delgado
Capable adversary's forces threatened?	**Yes**	**Yes**	**No**	**No**
PMCs tactically engaged?	**Yes**	**Yes**	**No**	**Yes**
Adversary partner force threatened?	**Yes**	**Yes**	**No**	**No**
Overt or covert Russian state military forces present in immediate area?[a]	No	Yes	Yes	No
Resource-related financial incentive?	Yes	No	Yes	Yes
Prior deployment of Russian PMCs in country?	Yes	Yes	No	No

NOTE: See coding justifications in Appendix G. Bold indicates variables most likely to produce escalation.

[a] We define *immediate area* as the location and vicinity to which Russian PMCs are deployed. Therefore, in the case of Deir Ezzor, we coded no for this variable because there were no Russian state forces near the battle involving the PMCs.

PMCs threatening an adversary's partner forces also might contribute to escalation; however, the significance of this variable is less clear than the two already discussed. PMCs might threaten an adversary's partner force, yet that adversary might not be willing to escalate if only its partner is threatened and not its own state forces. In both Deir Ezzor and Tripoli, PMC actions threatened a state adversary's uniformed military and partner forces. The evidence, therefore, does not allow us to identify an independent causal effect of PMC threats to partner forces in the absence of threats to an adversary's uniformed military.

Conversely, the presence of Russian state forces could limit escalation of an incident involving PMCs. In Deir Ezzor, U.S. military personnel earnestly attempted to deconflict with their Russian counterparts and confirm that Russian state service members were not among the forces assaulting the Conoco gas facility. On receiving reassurance that Russian state service members were not involved, U.S. forces responded decisively. Had Russian state service members been embedded with the assaulting forces, the U.S. response might have been more measured and designed to deescalate the incident with minimal casualties inflicted rather than the devastation of the

PMCs and pro-regime fighters that ensued. That said, Turkey does not seem to have been deterred by the Russian military's presence in Libya. Perhaps the covert nature of that presence minimizes its deterrent effect.

The prior deployment of Russian PMCs to an area and presence of resource-related incentives did not, by themselves, drive escalation in any of the four cases. However, both variables might provide insight into where and when Russian PMCs might be active in the future. Russian PMCs had previously deployed in both Syria and Libya, providing force protection and other security services. Such deployments might set the groundwork for other PMCs to deploy in the future. Additionally, Russian PMCs consistently deploy to environments where there are strong commercial incentives for some combination of the PMCs themselves (as in Deir Ezzor); Russian firms, particularly those associated with the PMCs (CAR); or the Russian state (Cabo Delgado). Therefore, previous deployment of Russian PMCs or the presence of commercial interests of Russian actors might be an indicator of a future deployment of Russian PMCs, perhaps engaged at the tactical level.

The cases of Syria, Libya, the CAR, and Cabo Delgado illustrate that Russian PMCs are likely to expand their activities in countries of geostrategic interest to Moscow, particularly states that hold sizable resource wealth and are coping with insurgencies or other forms of internal instability. Nigeria offers one possible future flashpoint scenario in this regard. The country holds the world's tenth largest proven oil reserves, it has grappled with the Boko Haram insurgency for over a decade, and the Nigerian government has indicated that Russia could support counterterrorism efforts. Additionally, in April 2018, then–U.S. Secretary of Defense James Mattis estimated that the United States maintained approximately 1,000 U.S. service members in the Sahel region to combat Salafi-jihadi insurgencies.[74] The possible deployment of Russian PMCs to Nigeria to conduct tactical counterinsurgency operations could put the PMCs in close proximity to U.S. forces, portending a possible flashpoint.

---

[74] Greg Myre, "The Military Doesn't Advertise It, but U.S. Troops Are All over Africa," NPR, April 28, 2018; and Nigerian Minister of Defense Mansur Dan Ali, quoted in Sergey Sukhankin, "The 'Hybrid' Role of Russian Mercenaries, PMCs and Irregulars in Moscow's Scramble for Africa," Jamestown Foundation, January 10, 2020a.

Conclusion

The foregoing analysis leads us to several observations and key takeaways:

- Despite the intense focus on Russian OIE in the press and among Western governments, none has led to an interstate conflict, and even the most plausibly escalatory of these operations has not come close to producing a flashpoint. This, in itself, is a critical and somewhat counterintuitive finding. Nonetheless, notable recent examples illustrate that such OIE will continue to be a key element of Russia's toolkit to advance its interests and thus merit continued attention in the future.
- An attributed attack that inflicted physical damage is probably much more likely to escalate to a flashpoint than one that does not cause damage or is not attributed in a timely fashion. These less escalatory attacks could be part of a broader spiral that leads to war, but the evidence suggests that such attacks on their own are less likely to result in flashpoints.
- The presence and tactical engagement of PMCs in a paramilitary role and their targeting of a capable adversary or that adversary's partner force appear to be the PMC-related variables most strongly correlated with escalation.
- Resource-related financial incentives and prior Russian PMC deployments in a country might indicate future areas of Russian PMC deployment.

CHAPTER SIX

Conclusions and Implications for the Army

In this report, we used a variety of analytical tools to better understand and anticipate flashpoints with Russia. We leveraged several different approaches in an effort to understand broader Russian conflict and dispute trends and to derive possible future flashpoints to which the U.S. military, and the Army in particular, might be called on to respond. We focused in particular on potential flashpoints with non-NATO allies. Even though this report was written before the full-scale invasion of Ukraine in 2022, its conclusions are all the more important in light of those events.

Our qualitative examination began by constructing a data set, building on and refining existing MIDs data, and categorizing the nature of past Russian run-ins with other states from 1992 to 2019. For this purpose, we distinguished between *flashpoints*, or militarized conflicts (defined as interstate clashes or interventions in civil war that result in combat deaths) and *militarized disputes* (interstate frictions that did not lead to recorded battle deaths). Drawing on the political science literature on escalation and the causes of conflict, we developed a 16-factor framework of possible drivers. We then used that framework to conduct in-depth case studies of four flashpoints, and did comparative analysis across those cases and against disputes with the same states that did not escalate to conflict.

We further leveraged quantitative analytical techniques to better understand the factors or correlates most associated with the onset of interstate frictions with Russia, using the yearly dyadic format of the MIDs data. This mode of analysis allowed us to also undertake a predictive exercise in an effort to gauge which countries, based on their characteristics, might be most prone to future MIDs with Russia. In addition to these lines of effort focused largely on traditional, overt interstate clashes or open interventions in civil wars, we examined additional potential drivers of conflict not captured in the historical data. Specifically, we analyzed Russia's use of PMCs and OIE to see whether either has the potential to lead to a flashpoint.

Findings

Both our qualitative and quantitative analyses reinforce the centrality of geographic proximity in driving Russian conflicts and disputes. All Russian militarized conflicts from 1992 to 2019 except the two related to the Syria operation have been located in the former Soviet region. The quantitative analysis underscored the correlation of territorial contiguity and former Soviet republic status with MID onset. The case studies further clarified the causal relationship between proximity to Russia and escalation.

That Russia has maintained a willingness to engage militarily in its so-called near abroad is not a revolutionary finding. However, it remains an important fact when considering whether and where Russia might become involved in militarized conflicts or disputes in the future, as Russia's 2022 invasion of Ukraine demonstrated. Russia shares a land border with 16 different countries—more than any other country in the world. This fact, coupled with the finding about territorial contiguity, suggests that the potential for Russian conflict spans many countries and several important regions of the world.

Beyond Russia's immediate neighborhood, the quantitative analysis in this report points to two additional factors correlated with Russian MIDs. First, states with unresolved land or maritime border issues with Russia are more likely to find themselves involved in a MID with Russia. Disagreements related to territory and the demarcation of national boundaries have historically often been the source of militarized disputes and engagements. Russia is no exception. As noted, Russia borders more countries than any other. By our accounting, Russia had ongoing territorial disagreements with six other states in 2019, but that number has been as high as 15 since 1992. Where there is ambiguity, uncertainty, or historical precedence for disagreement over borders, there is the potential for a clash.

Time is another factor to emerge from the quantitative models associated with Russian MID onset with other states. Simply put, a recent MID between Russia and another state strongly suggests that another MID might be in the offing. Russian engagement with Georgia, where it had multiple disputes in the several years leading up to the 2008 war is an apt example. This finding also suggests that the maxim "time heals all wounds" seems to apply to Russia's relations with other states.

The data also suggest that Russia has a history of MIDs with both major powers and smaller countries.[1] Looking forward, major powers might expect to find themselves entangled in disputes with Russia. How such engagements are handled will likely dictate their severity; our analysis does not necessarily presage escalation to the point of combat deaths. But the data do suggest that Russia has not shied away from MIDs in the past with major powers.

Our qualitative analysis highlights the role of broader geopolitical factors in driving conflict escalation with Russia. Russia's dissatisfaction with its place in the international system, acute uncertainty about the future, and reputational costs were central to the escalatory dynamic in all four conflicts examined in the case studies of militarized conflict. This was true for both interstate clashes and interventions in civil wars. However, external threats to Russia were often the immediate trigger. External security threats were identified as drivers of escalation in all of the conflicts. Taken together, these observations reinforce the view of Russia as a status-seeking, geopolitically minded, but predominately regional power—or at least a power that sees its immediate environs as the primary source of security threats. Furthermore, the centrality of external threats to Russia's calculus across the cases suggests that Russia is driven to war by the perception of imminent potential losses as much as by an expansionist instinct.

Applying the same framework to test for the presence of these potential drivers in cases of disputes with Ukraine and Georgia that did not escalate to conflict reinforced the importance of two of the geopolitical factors: dis-

---

[1] It is worth noting that our quantitative analysis looked specifically to identify factors associated with Russian propensity to engage in a MID, which refers both to episodes that did escalate to conflict and those that did not. (It also excludes interventions against nonstate actors, such as Russia's operation in Syria).

satisfaction with the international systemic status quo and acute uncertainty about the future. It also underscored the importance of external threats. These three factors were absent when disputes with the same countries over broadly similar issues did not escalate to the point of a militarized conflict, and were present when they did.

Two additional, though more tenuous, results emerge from the qualitative analysis. Both opposing states in the interstate militarized conflicts that we examined were considered democratic at the time, but we could not identify a causal role that regime type played in driving escalation in either case. This finding calls into question the commonly encountered notion that Russia is driven to war to prevent democracy from emerging on its periphery. Second, all nine identified militarized conflicts took place in states where Russia had a military installation. We had assumed that the presence of a Russian military facility in the state where a conflict was located could drive escalation by facilitating Moscow's intervention. However, in those cases for which the presence of a facility was an important driver of escalation, they were more of a liability, entangling Russia in a conflict because its forces came under threat, rather than an asset that allowed for enhanced power projection beyond its borders. Even Russia's alliance obligations only played into *disputes* when its own forces come under threat.

Finally, our examination of two additional potential drivers of conflict—OIE and the use of PMCs—reveals several important dynamics. A first finding is somewhat counterintuitive: There were no cases of OIE that led to an interstate conflict, and even the most plausibly escalatory of these operations have not come close to producing a flashpoint. We document two such operations, a cyberattack on a Saudi oil refinery and an electromagnetic attack on military and civilian assets during a NATO exercise in Northern Europe, that arguably came the closest to escalating. These examples demonstrate that such OIE can have significant strategic effects and thus merit continued attention in the future. The analysis suggests that the characteristics of OIE most likely to escalate to a flashpoint are an attributed attack that inflicted physical damage.

Russian use of PMCs in zones of conflict or unstable countries has the potential to ignite an interstate clash. Certainly, the 2018 Deir Ezzor incident in Syria came precipitously close to producing such an outcome. The tactical engagement of PMCs in a paramilitary role and their targeting of a capable adversary or partner force appear to be the PMC-related variables most strongly correlated with escalatory dynamics.

In summary, the qualitative drivers of escalation to conflict in the period we examined are as follows:

- proximity
- geopolitical drivers
  - dissatisfaction with its place in the international system
  - acute uncertainty about the future
  - reputational costs
- external threats.

The quantitative drivers of MID onset are as follows:

- contiguity
- territorial disagreements
- former soviet states
- real oil rents (Russia)
- time (reverse correlation)
- major-power status.

The characteristics of the additional factors that could create escalatory dynamics are as follows:

- OIE: timely attribution, physical damage
- PMCs: tactical engagement, threatening capable adversary or adversary's partner forces.

Implications

The implications of this project for the Army do not follow in straightforward manner from the research findings. We have examined Russia's history of conflicts and disputes from 1992 to 2019 and what it might portend; we do not prescribe U.S. responses to potential future scenarios. That said, there are a variety of implications for the Army that emerge from the research presented here. First, the Army might be called on to conduct or complement operations to counter Russian flashpoints with countries not currently central to Army planning considerations. The full-scale Russian invasion of Ukraine in 2022 was a vivid demonstration.

A second implication concerns the frequency and propensity for Russia to engage in both militarized conflicts and disputes. Militarized conflicts involving Russia were relatively rare from 1992 to 2019, and have not involved major powers to date.[2] Militarized disputes, however, were not infrequent. Although our data set on disputes ends in 2010, anecdotal evidence suggests that they might have been even more common in recent years.

A third key implication stems from our finding about the centrality of proximity in driving escalation. Army engagements with countries in Russia's immediate periphery in post-Soviet Eurasia could therefore be a source of a dispute or even conflict involving Russia. Appreciating this potential in advance can help mitigate future incidents and possibly prevent escalation.

Finally, such Russian activities as the deployment of PMCs and use of OIE are problematic but have not produced interstate conflict to date. Although the Deir Ezzor incident indicates that PMC tactical operations can lead to U.S. military engagements with these parastatal actors, even that case did not become an interstate clash. Moreover, it is not obvious that the United States, and the Army in particular, should or need to respond in kind to every Russian provocation involving operations in the information environment or the use of PMCs. Publicly highlighting or revealing Russian hostile or covert activities might be a better way to counteract such behavior.

Recommendations

Planners cannot possibly take into account every possible contingency and consideration when thinking about how the Army might be called on to respond to a flashpoint scenario with Russia. However, a few key drivers of escalation in Russia-related contingencies should certainly inform and motivate planning. Territorial contiguity with Russia, former Soviet republic status, and unresolved border issues are characteristics of states likely to be

---

[2]   There were two cases of conflict with Turkey, but both involved a single combat death and were short-lived.

engaged in disputes with Russia. The full-scale Russian invasion of Ukraine in 2022 reinforced the validity of this finding. Relatedly, the Army might expect that recent disputes might serve as an augur for future ones, and possibly flashpoints. Finally, the broader geopolitical context at the time of any interaction with Russia should be considered when gauging the risk of conflict. Whatever its actual weight in the international system, Russia acts like a great power: Geopolitical considerations can drive its decisionmaking about war and peace.

Another central recommendation to emerge from our research deals with planning for the *expected surprise*. The analysis points to several states where flashpoints are possible, not all of which are likely at the top of the Army's planning agenda. Future flashpoint scenarios might very well take the Army by surprise. The degree to which Army planning is able to take this into account should increase its capacity to forge a timely and effective response.

Looking forward, the Army will want to consider how it might be asked to respond to various potential future flashpoints. As already noted, future scenarios call for their own analysis and associated wargaming efforts. This is the best way to ensure that conflict contingencies are met with as coordinated a response as possible. Factors the Army will want to take into account are force planning (including manpower and readiness), posture requirements for both short and more-sustained engagements, partner relationships, security assistance arrangements, sustainment efforts, and the challenges associated with the movement of troops.

Russia does not shy away from militarized disputes, particularly on and near its borders, and has occasionally been engaged in interactions that escalated to the level of a flashpoint. It also resorts to unconventional methods of influence, such as the use of PMCs or engaging in cyber activities. These behaviors need not precipitate a conflict to which the Army is asked to respond. But they certainly could. The more the Army readies itself to deal with potential flashpoints, particularly in areas outside its alliance responsibilities, the easier the task will be should it ever present itself.

Russia's Militarized Disputes, 1992–2010

This table lists the 54 militarized disputes in our data set, which is derived from the MIDs data. We created the dispute categories for analytical purposes. These categories and the disputes themselves are discussed in detail in Chapter Two.

TABLE A.1

Russia's Militarized Disputes, 1992–2010

Dispute Number	Opposing State	Location	Start Date	Dispute Category	MID Number
1	Ukraine	Sevastopol, Ukraine	July 1992	Defense of or response to attacks on Russian forces or assets abroad	3559
2	Estonia	Estonia-Russia Border	July 1992	Defense of or response to attacks on Russian forces or assets abroad	3560
3	Sweden	Sweden territorial waters	September 1992	Signaling or deterrence	3563
4	Moldova	Dniestr River	February 1993	Protracted regional conflict	4051
5	China	China Exclusive Economic Zone	June 1993	Defense of or response to attacks on Russian forces or assets abroad	4052

Table A.1—Continued

Dispute Number	Opposing State	Location	Start Date	Dispute Category	MID Number
6	Poland	Sea of Okhotsk	June 1993	Commercial- or fishing-related	4106
7	Japan	Japan-Russia border (territorial waters)	November 1993	Border-related	4042
8	Latvia	Riga, Latvia	January 1994	Defense of or response to attacks on Russian forces or assets abroad	4103
9	Ukraine	Odessa, Ukraine	April 1994	Defense of or response to attacks on Russian forces or assets abroad	4050
10	Afghanistan	Afghanistan-Tajikistan border	December 1994	Defense of or response to attacks on Russian forces or assets abroad	4055
11	Lithuania	Vilnius, Lithuania	March 1995	Other	4105
12	Ukraine	Rostov-on-Don, Russia	March 1996	Signaling or deterrence	4098
13	Turkey	Armenia-Turkey Border	July 1996	Defense of or response to attacks on Russian forces or assets abroad	4173
14	China	Amur River	October 1996	Border-related	4104
15	Japan	Nemuro Strait	October 1996	Border-related	4297
16	Poland	Sea of Okhotsk	February 1997	Commercial- or fishing-related	4107
17	Afghanistan	Afghanistan-Uzbekistan border	May 1997	Defense of or response to attacks on Russian forces or assets abroad	4176

Table A.1—Continued

Dispute Number	Opposing State	Location	Start Date	Dispute Category	MID Number
18	United States	Bering Sea	August 1997	Commercial- or fishing-related	4174
19	Georgia	Zemo Larsi, Georgia	August 1997	Other	4096
20	Norway	Waters off of Svalbard/ Spitsbergen	July 1998	Commercial- or fishing-related	4321
21	Latvia	Latvia-Russia border	August 1998	Border-related	4111
22	Afghanistan	Afghanistan-Uzbekistan border	August 1998	Defense of or response to attacks on Russian forces or assets abroad	4228
23	Azerbaijan	Baku, Azerbaijan	March 1999	Defense of or response to attacks on Russian forces or assets abroad	4338
24	NATO states	Adriatic Sea	April 1999	Signaling or deterrence	4342
25	Turkey	Turkey	April 1999	Other	4344
26	United Kingdom	Kosovo	June 1999	Signaling or deterrence	4334
27	Norway	Norway territorial waters	June 1999	Signaling or deterrence	4335
28	Afghanistan	Afghanistan-Tajikistan border	September 1999	Defense of or response to attacks on Russian forces or assets abroad	4201
29	Georgia	Shatili, Georgia	November, 1999	Other	4212
30	United States	Gulf of Oman	February 2000	Defense of or response to attacks on Russian forces or assets abroad	4213
31	Japan	Hokkaido Island, Japan	April 2000	Border-related	4222

Table A.1—Continued

Dispute Number	Opposing State	Location	Start Date	Dispute Category	MID Number
32	United States	Sea of Japan	October 2000	Signaling or deterrence	4220
33	Canada United States	Arctic regions of Russia	November 2000	Signaling or deterrence	4197
34	Japan	Hokkaido Island, Japan	February 2001	Signaling or deterrence	4239
35	Norway	Norway territorial waters	February 2001	Signaling or deterrence	4238
36	Georgia	Georgia-Russia border	October 2001	Other	4242
37	Georgia; Azerbaijan	Georgia-Russia border and Azerbaijan-Russia border	May 2002	Signaling or deterrence	4411
38	Argentina	Argentina territorial waters	August 2002	Commercial- or fishing-related	4494
39	Georgia	Tskhinvali, South Ossetia, Georgia	February 2003	Protracted regional conflict	4416
40	Denmark	Baltic Sea	April 2003	Signaling or deterrence	4417
41	Sudan	Sudan	July 2003	Defense of or response to attacks on Russian forces or assets abroad	4360
42	Georgia	Georgia-Russia border	September 2003	Protracted regional conflict	4420
43	Georgia	Georgia	April 2004	Protracted regional conflict	4422
44	Georgia	Georgia-Russia border	March 2005	Protracted regional conflict	4424
45	Ukraine	Feodosiya, Crimea, Ukraine	March 2005	Other	4425

Table A.1—Continued

Dispute Number	Opposing State	Location	Start Date	Dispute Category	MID Number
46	Norway	Norway-Russia border (territorial waters)	October 2005	Commercial- or fishing-related	4429
47	Japan	Japan-Russia border (airspace)	January 2006	Signaling or deterrence	4475
48	Japan	Habomai, Southern Kurils/Northern Territories	August 2006	Border-related	4477
49	Finland	Finland-Russia border (airspace)	December 2007	Signaling or deterrence	4437
50	Japan	Izu Islands, Japan	February 2008	Signaling or deterrence	4480
51	Ukraine	Sevastopol, Ukraine	April 2008	Other	4438
52	Norway	Waters off of Svalbard/ Spitsbergen	July 2008	Commercial- or fishing-related	4440
53	Japan	Kuril Islands	January 2009	Border-related	4484
54	China	Russia territorial waters	February 2009	Commercial- or fishing-related	4485

Figure A.1 maps the disputes listed in Table A.1. The opposing states in these disputes are colored darker blue depending on the number of Russian disputes they experienced.

FIGURE A.1

Heat Map of the Opposing State in Disputes with Russia, 1992–2010

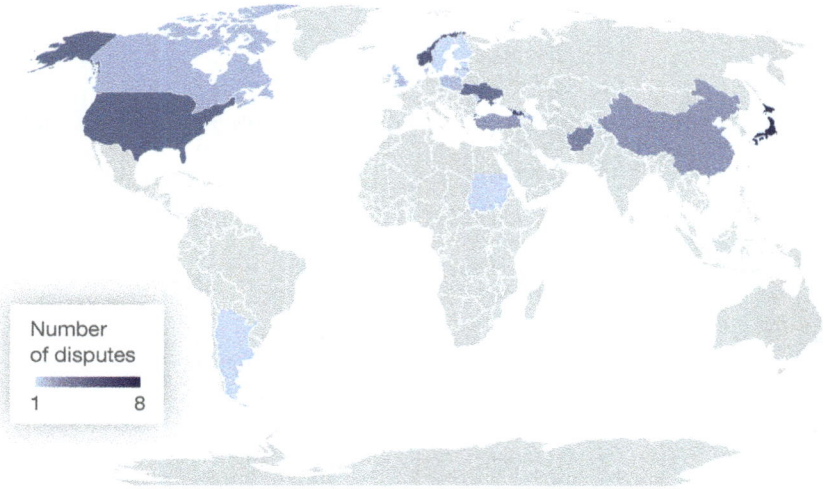

Militarized Conflict Case Study: Ukraine Intervention, 2014

In late November 2013, Ukraine's then-president Viktor Yanukovych abruptly halted his country's ongoing negotiations with European officials over Ukraine's Association Agreement (AA) with the EU.[1] This move, which was perceived as a signal of the demise of Ukraine's lengthy journey toward closer partnership with the EU, provoked major antigovernment unrest in Ukraine, a violent response by Ukrainian authorities, the departure of the Yanukovych regime, and the ushering in of a new, more pro-Western (and anti-Russian) government. With the replacement of Kremlin-friendly Yanukovych by new pro-Western officials in Kyiv, closer integration with western Europe and distancing from Moscow appeared all the more likely. Although this development was welcomed by many in Ukraine and the West, it likely triggered Kremlin anxieties and sparked unease among some Russophone and pro-Russian groups in Ukraine.[2]

In early March 2014, amid the transition of power to Kyiv's new interim government, Russian forces seized Crimea and ultimately annexed the peninsula.[3] On the heels of Russia's incursion into Crimea, protests and unrest aimed at the new sitting government in Kyiv erupted in the Donbas macroregion of eastern Ukraine, which is the focus of this case. Activists calling for independence from Ukraine, some of whom carried Russian flags

---

[1] Ian Traynor and Oksana Grytsenko, "Ukraine Suspends Talks on EU Trade Pact as Putin Wins Tug of War," *The Guardian*, November 21, 2013.

[2] Paul D'Anieri, *Ukraine and Russia: From Civilized Divorce to Uncivil War*, Cambridge, United Kingdom: Cambridge University Press, 2019.

[3] D'Anieri, 2019, pp. 228–229.

and requested Russian support, occupied government buildings in Donetsk, Luhansk, and Kharkiv in early April 2014 and announced the establishment of the Donetsk, Luhansk, and Kharkiv People's Republics, respectively.[4]

Spurred by additional separatist violence, authorities in Kyiv announced the launch of an "Anti-Terrorist Operation" (ATO) against separatist elements in eastern Ukraine on April 13, after which fighting between Ukrainian government forces and separatists in the Donbas region escalated.[5] As part of the Ukrainian counteroffensive to reassert control in the region that day, a clash between Ukrainian security services and pro-Russian separatists in the eastern Ukrainian city of Slavyansk reportedly resulted in fatalities on both sides.[6] Evidence indicates that these were the first casualties to occur as a result of fighting between Ukrainian forces and pro-Russian separatists in eastern Ukraine;[7] using this project's definition of militarized conflict (one or more incidents of Russia's use of force toward another state—or vice versa—that result in at least one battle death), these casualties represent the point at which the dispute in the Donbas region escalated to militarized conflict.

In this appendix, we focus on identifying and examining the drivers that motivated this specific escalation to conflict in the Donbas between Ukrainian and Russian forces. Recognizing that the extent of Russian involvement at this phase of the conflict is debated, we nonetheless classify the separatist pro-Russian forces embroiled in this event as Russian forces.[8] Western officials have pointed to evidence that they assert implicates Russian troops in

---

[4] Richard Balmforth and Lina Kushch, "Pro-Moscow Protesters Seize Arms, Declare Republic, Kiev Fears Invasion," Reuters, April 7, 2014; and David M. Herszenhorn and Andrew Roth, "In East Ukraine, Protesters Seek Russian Troops," *New York Times*, April 7, 2014.

[5] D'Anieri, 2019, p. 234.

[6] Andrew E. Kramer and Andrew Higgins, "Ukraine Forces Storm a Town, Defying Russia," *New York Times*, April 13, 2014; and "Ukraine Takes Action in Slovyansk; West Condemns Russian Involvement," Radio Free Europe/Radio Liberty, April 13, 2014.

[7] "Clashes in Eastern Ukraine Reportedly Turn Deadly," NPR, April 13, 2014; and "Ukrainian Clashes with Pro-Russian Separatists Turn Deadly," *Washington Post*, April 13, 2014.

[8] D'Anieri, 2019, p. 233; Charap and Colton, 2017, pp. 131–132.

the Slavyansk clashes on April 13. For instance, Anders Fogh Rasmussen, then the secretary general of NATO, identified "specialized Russian weapons and identical uniforms without insignia previously worn by Russian troops during Russia's illegal and illegitimate seizure of Crimea" as proof that Russian troops were present at Slavyansk.[9] The U.S. Department of State cited similar evidence and referenced information demonstrating that Russian intelligence had a hand in organizing the separatist efforts.[10] Comments made by Samantha Power, then the U.S. ambassador to the United Nations, echoed these assertions. She described the pro-Russian separatist activities in eastern Ukraine at the time as "professional" and "coordinated" and as "bear[ing] the tell-tale signs of Moscow's involvement."[11] Russian officials, however, have consistently maintained that no Russian military forces have been involved in the fighting.

Although the evidence placing Russian troops in the Donbas at the time of conflict escalation is not infallible, it does imply some level of Russian support of the separatist forces. What is more, Russian involvement in the conflict became more evident over the course of the summer. By August 2014, Ukrainian forces, who were more organized and capable by this point, appeared poised to deliver a decisive blow to separatist forces near the Ukrainian city of Ilovaisk.[12] But in the Battle of Ilovaisk, Ukrainian forces ultimately suffered major casualties at the hands of separatists and regular Russian troops whose presence has since been substantiated by evidence.[13]

Several additional characteristics of the conflict in the Donbas make it particularly challenging to tease out factors that contributed to escalation in fighting between Ukrainian and Russian forces. First, the fighting in eastern Ukraine involved more than two state-level parties engaged in conventional military combat. The Ukrainian side involved both official

---

[9] "Ukraine Crisis: Casualties in Sloviansk Gun Battles," *BBC News*, April 13, 2014.

[10] Paul Lewis, "UN Ambassador: Ukraine Unrest Has 'Tell-Tale Signs of Moscow's Involvement,'" *The Guardian*, April 13, 2014.

[11] Lewis, 2014.

[12] D'Anieri, 2019, p. 245.

[13] Oksana Grytsenko, "Thousands of Russian Soldiers Fought at Ilovaisk, Around a Hundred Were Killed," *Kyiv Post*, April 6, 2018; and Shaun Walker, "New Evidence Emerges of Russian Role in Ukraine Conflict," *The Guardian*, August 18, 2019.

government forces and volunteer forces, sometimes referred to as volunteer battalions or militias.[14] These forces are believed to have fought against a mosaic of Russian-supported Ukrainian separatists, Russian-sponsored mercenaries (such as the now infamous Wagner Group and foreign fighters) and uniformed Russian military forces.[15] Additionally, the Russian government obscured its role in the Donbas, and Russian officials consistently denied the presence of regular Russian forces in eastern Ukraine.[16] As noted already, this opacity increased the difficulty in determining the extent to which Russia was involved in any specific escalatory event and the extent to which these events can be attributed to purposeful Russian provocation at any given time.

Lastly, we deliberately exclude Russia's invasion and annexation of Crimea from our analysis. Given that Russian military efforts in the peninsula resulted in no combat deaths, the Crimean case did not meet our definition of a militarized conflict (which entails at least one combat-related death) within this project's framework.

---

[14] These volunteer forces were composed of a patchwork of individuals made up of retired Ukrainian law enforcement, Ukrainian activists from the Euromaidan, far-right nationalists, foreign nationals, and others. Some have since been formally subordinated under Ukrainian government command and control. For additional details, see Maria Antonova, "They Came to Fight for Ukraine. Now They're Stuck in No Man's Land," *Foreign Policy*, October 19, 2015; Patrick Jackson, "Ukraine War Pulls In Foreign Fighters," *BBC News*, September 1, 2014; and Tetyana Malyarenko and David J. Galbreath, "Paramilitary Motivation in Ukraine: Beyond Integration and Abolition," *Southeast European and Black Sea Studies*, Vol. 16, No. 1, 2016.

[15] Walker, 2019. The U.S. Treasury Department has sanctioned Russian mercenary groups, such as the Wagner Group, for involvement in fomenting the conflict in the Donbas. See, for example, U.S. Department of the Treasury, "Treasury Designates Individuals and Entities Involved in Ongoing Conflict in Ukraine," press release, Washington, D.C., June 20, 2017.

[16] "Putin Claims Russia Was 'Forced to Defend Russian-Speaking Population in Donbass,'" *The Interpreter*, October 12, 2016.

Drivers

Our analysis indicates that there are 11 drivers of escalation present in this case:

1. shared land border between Ukraine and Russia
2. Ukraine's status as a former Soviet republic with shared Soviet legacy
3. the salience of the Donbas for Russia
4. the presence of Russian compatriots in the Donbas region
5. a power preponderance in Russia's favor
6. Russian dissatisfaction with its role in the international system
7. Russian perceptions of increased external security threats
8. Russia's acute uncertainty about the future
9. reputational costs for Russia for nonintervention
10. domestic instability in Ukraine
11. the Donbas represented a critical territorial issue for Ukraine.

These drivers of escalation in the Donbas conflict can be grouped into two broad categories: structural triggers and immediate or proximate triggers. Such factors as the geographic proximity of Ukraine and Russia (i.e., shared land border and former Soviet republic status), the economic and military salience of the Donbas for Russia, the existence of Russian compatriots in the Donbas, the relative economic and military strength of Russia over Ukraine, and Russian dissatisfaction with the country's position in the international system all contributed to the fraught relationship between Moscow and Kyiv.[17] In this sense, these longer-term structural factors were necessary in establishing the preconditions for conflict between Ukraine and Russia. That said, these factors were not sufficient: All had been in place for some time but had not resulted in conflict between Ukraine and Russia prior to 2014.

This previous stasis indicates that other, more-acute, proximate factors were responsible for escalating the crisis between Moscow and Kyiv into a militarized conflict. As we discuss in detail later in this appendix, these immediate triggers consisted of heightened Russian threat perceptions,

---

[17] D'Anieri, 2019, p. 2.

acute Russian uncertainty about the future, Ukrainian domestic political instability, major threats to Ukrainian territorial integrity, and the prospect of damaging reputational costs to Russia for nonintervention.

The Ukraine case demonstrates that Russia was willing to covertly use militarized force to protect its interests in Ukraine only when the Kremlin felt those interests were acutely threatened. That said, Russian decision-makers' unwillingness to acknowledge the presence of Russian forces in the Donbas indicates that they were unwilling to incur the potential costs and risks associated with an overt conventional incursion into and annexation of eastern Ukraine. Our analysis also found that the mere presence of Russian compatriots in the Donbas in itself was not the determining driver of escalation but did contribute to the existing dynamics on the ground. Rather, Russia's decision to escalate likely arose more from the expected reputational costs of violating the Kremlin's (and Putin's) public assurances that compatriots would be protected. Lastly, the Kremlin might have been less willing to escalate to conflict if it had viewed the Ukrainian government, economy, or military as strong, though the disparity in power was not a proximate factor, as we will explain.

This case also demonstrates that acute challenges to Ukraine's territorial integrity played an important role in the escalation. Ukrainian authorities' perceptions that the country's territorial integrity was at risk, heightened by the earlier demonstration in Crimea that the Kremlin was willing and able to mount a successful incursion into Ukraine and annex internationally recognized Ukrainian territory, motivated Ukraine's decision to launch a counteroffensive that contributed to escalation.

Geographic and Territorial Drivers

Geographic Drivers: Shared Land Border and Former Soviet Republic Status

Ukraine and Russia share a long, complex, and storied history.[18] The two states also share a land border that is longer than 1,400 miles (or 2,200 km).[19]

---

[18] For a detailed discussion of Ukraine and Russia's history, see D'Anieri, 2019.

[19] Elias Götz, "It's Geopolitics, Stupid: Explaining Russia's Ukraine Policy," *Global Affairs*, Vol. 1, No. 1, 2015, p. 3.

The Kremlin is a mere 8.5-hour car ride away from Ukraine's eastern city of Sumy. The Ukrainian-Russian border also lies close to Russia's Volga region, the "industrial and political heartland of the Russian Federation."[20] The geographic proximity and respective locations of Ukraine and Russia are significant in the context of the Donbas conflict for two interconnected reasons: geopolitical considerations and the states' shared historical, ideational, and cultural links resulting from Ukraine's status as a former Soviet republic.

From Moscow's vantage point in 2014 (and historically speaking as well), Ukraine was situated in an area of extreme strategic importance and insecurity for Russia. Russian decisionmakers and their Soviet predecessors have long conceived of Ukraine as an integral buffer between Russia and Western Europe. According to scholar Andrei Tsygankov, "the ethno-territorial borders of Ukraine were drawn by Joseph Stalin . . . in order to establish a meaningful buffer zone protecting the Soviet Union from the West."[21] Despite the dissolution of the Soviet Union and Ukraine's independence, many Russian policymakers continue to view Ukraine through this lens.[22] This concern has become all the more acute in the post–Cold War era as Western political and military institutions, notably the EU and NATO, enlarged eastward toward Russia. From the Kremlin's perspective, EU and NATO accession of states that Russia traditionally considered to be within its orbit meant that Ukraine became the frontier between Russia and Western political and defense alliances.[23]

Thus, when it became clear that the government that came to power after the Maidan Revolution was committed to pursuing the EU AA, Russian insecurities related to Ukraine's geographic location and its strategic sig-

---

[20] Götz, 2015, p. 3.

[21] Andrei Tsygankov, "Vladimir Putin's Last Stand: The Sources of Russia's Ukraine policy," *Post-Soviet Affairs*, Vol. 31, No. 4, February 2015, p. 288.

[22] Tsygankov, 2015, p. 288.

[23] After NATO publicly stated in 2008 that Ukraine and Georgia would eventually join the Alliance, President Putin reaffirmed previous statements that Russia would consider NATO enlargement to Ukraine or Georgia as a "direct threat." See Anil Dawar, "Putin Warns NATO over Expansion," *The Guardian*, April 4, 2008.

nificance for Russia were heightened.[24] As scholar Elias Götz points out, the AA was not purely economic in nature, and it included provisions for the integration of Ukrainian and European defense and security policy.[25] Likewise, Götz asserts that many Russian decisionmakers viewed EU agreements as preludes to NATO membership. Through the Kremlin's lens, NATO membership for Ukraine would place the military troops, equipment, and installations of its primary adversary adjacent to Russia's border, which would likely trigger most states' insecurities that found themselves in this position.[26]

Ukrainians and Russians both associate the founding of their nations (and states) with the Kyivan Rus', a federation of peoples that existed from the late ninth to 13th centuries in what is now modern-day Ukraine, Belarus, and areas of western Russia.[27] Likewise, both Russians and Ukrainians claim Kyiv in particular as the birthplace of their respective nations. Starting in the 17th century, the territory that makes up Ukraine today came under the rule of various external states and empires, including Poland and the Russian empire.[28] Those parts of eastern Ukraine that were under Russian tsarist rule at the time of that regime's demise were incorporated into the Soviet Union first. Other areas of Ukraine came under Soviet control during the Second World War.[29] This is a highly simplified rendition of this history, but its contours are integral to Ukrainian and Russian identities, threat perceptions, and politics. Broadly speaking, Ukrainian oppression at the hands of Russian and Soviet rulers is an important element of Ukrainian national identity.[30] On the other hand, many Russians point to this shared

---

[24] Götz, 2015.

[25] Götz, 2015, p. 4.

[26] Götz, 2015, p. 5; Tsygankov, 2015, p. 288.

[27] Rajan Menon and Eugene Rumer, *Conflict in Ukraine: The Unwinding of the Post-Cold War Order*, Boston: MIT Press, 2015, p. 4.

[28] Menon and Rumer, 2015, p. 5.

[29] Menon and Rumer, 2015, pp. 5–7.

[30] Karina Korostelina, "Shaping Unpredictable Past: National Identity and History Education in Ukraine," *National Identities*, Vol. 13, No. 1, March 2011.

history as evidence that Ukraine is an extension of Russia rather than an independent nation, culture, or state.[31] As scholar Paul D'Anieri notes:

> To many Russians, Ukraine is part of Russia, without which, Russia is incomplete. This belief is rooted in hundreds of years in which much of Ukraine was part of the Russian empire and Soviet Union, in the Russian foundation myth which sees the origins of today's Russia in medieval Kyiv, and in the important role played by people from Ukraine—Gogol, Trotsky, Bulgakov, and Brezhnev among many others—in Russian/Soviet culture and politics.[32]

It is possible that the Ukrainian authorities in Kyiv following Yanukovych's departure were primed to view Russian behavior through a threatening prism, a perspective that was only reinforced by Russia's violation of Ukraine's territorial integrity with its annexation of Crimea. For those Russian decisionmakers and elites who perceived Ukraine as an extension of Russia, the loss of Ukraine to the West would mean abandoning a piece of Russian culture, history, and identity.[33] These perceptions might have made both parties more prone to escalate.

As we will discuss in the following sections on acute uncertainty and security threats, Russia and Ukraine had long shared a border without engaging in conflict. The relationship between Kyiv and Moscow has been fraught with tension, but this tension did not escalate to conflict until the war in the Donbas in spring 2014. So, the deep-rooted and long-standing frictions between Kyiv and Moscow underpinning the conflict in the Donbas are partly attributed to these states' specific geographic location and physical proximity,[34] but the two geographic factors discussed in this section—shared land border and Ukraine's status as a former Soviet republic—did

---

[31] E. Wayne Merry, "The Origins of Russia's War in Ukraine: The Clash of Russian and European 'Civilizational Choices' for Ukraine," in Elizabeth A. Wood, William E. Pomeranz, E. Wayne Merry, and Maxim Trudolyubov, *Roots of Russia's War in Ukraine*, Washington, D.C.: Woodrow Wilson Center Press, 2015, pp. 37–38; and D'Anieri, 2019, p. 10.

[32] D'Anieri, 2019, p. 10.

[33] D'Anieri, 2019, pp. 10–12; Merry, 2015, p. 37; and Serhy Yekelchyk, *The Conflict in Ukraine: What Everyone Needs to Know*, New York: Oxford University Press, 2015, p. 6.

[34] Menon and Rumer, 2015, p. 2.

not by themselves escalate the conflict between Ukraine and Russia in the Donbas.

Territorial Drivers: Territorial Issue for Ukraine and Salience of the Donbas for Russia

The conflict in the Donbas was an important territorial issue for Ukraine; Ukrainian territorial integrity was directly threatened as a result of the hostilities. There are several reasons why the authorities in Kyiv perceived separatist activities in eastern Ukraine as a threat to the country's territorial integrity. First, the separatist attitudes that took hold after the collapse of the Yanukovych regime were real and manifested themselves in public protests.[35] Separatists demanded the independence of large swaths of Ukrainian territory, displayed such symbols as the Russian flag, and publicly appealed to the Kremlin for help.[36] Additionally, it was very important for Ukraine to maintain control over the territory under dispute because it is rich in natural resources and because it is home to a significant portion of the country's industry, including coal, steel, and other production.[37] The timing of these separatist efforts in the Donbas was also significant, coming on the heels of Russia's annexation of Crimea in March 2014. Moreover, events in Donbas bore some of the hallmarks of the Crimean case, such as the seizure of local administrative buildings.[38]

Remarks by the interim authorities in Kyiv indicate that Ukrainian decisionmakers perceived separatist activities in the Donbas as a pretext to another Crimea-like operation.[39] Interim president Oleksandr Turchynov

---

[35] See polling cited in Yuliya Mostovaya, Sergei Rakhmanin, and Inna Vedernikova, "Yugo-Vostok: vetv' dreva nashego," *Zerkalo nedeli*, April 18, 2014.

[36] Sam Frizell, "Violence in East Ukraine Ratchets Up Tensions with Russia," *TIME*, April 6, 2014; and Lina Kushch, "Pro-Russia Protesters Occupy Regional Government in Ukraine's Donetsk," Reuters, March 3, 2014.

[37] Jeanette Seiffert, "The Significance of the Donbas," *DW*, April 15, 2014; Kushch, 2014; and Zbigniew Wojnowski, "Economic Tensions Worsen Unrest in Eastern Ukraine," *Al Jazeera America*, March 25, 2014.

[38] Frizell, 2014; Kushch, 2014.

[39] Balmforth and Kushch, 2014; and Lina Kushch and Thomas Grove, "Pro-Russia Protesters Seize Ukraine Buildings, Kiev Blames Putin," Reuters, April 6, 2014.

described separatist activities in the Donbas in the first week of April as attempts to "dismember" Ukraine.[40] "Enemies of Ukraine," Turchynov said, were "trying to play out the Crimean scenario."[41] The interim prime minister Arseniy Yatsenyuk's comments echoed similar concerns. The "[Kremlin's] plan is to destabilize the situation, the plan is for foreign troops to cross the border and seize the country's territory, which we will not allow," Yatsenyuk asserted in an emergency cabinet meeting held in response to separatist activities in eastern Ukraine.[42] Russia had also begun massing thousands of troops on the Ukrainian-Russian border, and Russian President Putin had publicly stated that Russia would protect Russian speakers in Ukraine.[43] Given these developments, the authorities in Kyiv perceived Ukrainian territory to be imperiled.

Evidence indicates that this factor contributed to escalation. It was only after unrest in the Donbas appeared to seriously threaten Ukrainian territorial integrity that authorities in Kyiv declared the ATO and deployed government forces in a combat role, which contributed to the conflict's initial escalation.

A second factor related to territory, the salience of the disputed territory to Russia, also contributed to the drive to escalation in the Donbas conflict. The disputed territory (in this case, the Donbas) held important military, economic, and symbolic value for Russia. Eastern Ukraine was home to manufacturing that was important for the Russian defense sector and, thus, the Russian military.[44] Ukraine was the sole producer of some Russian

---

[40] Matt Smith and Victoria Butenko, "Ukraine Says It Retakes Building Seized by Protesters," *CNN*, April 7, 2014.

[41] Smith and Butenko, 2014.

[42] Danielle Wiener-Bronner, "Another Ukrainian City Wants Its Independence," *The Atlantic*, April 7, 2014.

[43] Adam Entous and Julian E. Barnes, "Russian Buildup Stokes Worries: Pentagon Alarmed as Troops Mass near Ukrainian Border," *Wall Street Journal*, March 28, 2014; and President of Russia, "Transcript: Putin Says Russia Will Protect the Rights of Russians Abroad," *Washington Post*, March 18, 2014b.

[44] Seiffert, 2014.

defense-related parts at the time.[45] Ukrainian-manufactured parts were particularly important for Russia's strategic nuclear forces.[46] For instance, as of July 2014,

> Ukrainian specialists carr[ied] out regular inspections of Russia's strategic missiles to certify them for service and supplying essential missile components including targeting and control systems for the RS-20 missile (known by NATO as the SS-18 Satan).[47]

Nature of the Relationship Between Russia and the Opposing State

Presence of Russian Military Installation

The presence of a Russian military installation in the territory of the state at the outset of the conflict poses a somewhat complex question. Crimea—notably the naval base at Sevastopol (which has been home to the Russian, and previously Soviet, Black Sea Fleet—has historically been perceived by Moscow as strategically significant for Russia's security and its military.[48] Even though Moscow had annexed Crimea by the time the conflict in question began, all but Russia and a few other states considered it sovereign Ukrainian territory. For this reason, and because Moscow could have used the facility for the invasion of the Donbas, it is our position that Russia did have a military installation on Ukrainian territory at the time of the conflict. However, there is no evidence that the base played any role in driving escalation of the conflict.

---

[45] Alexandra McLees and Eugene Rumer, "Saving Ukraine's Defense Industry," Carnegie Endowment for International Peace, July 30, 2014.

[46] McLees and Rumer, 2014.

[47] McLees and Rumer, 2014.

[48] Alan Yuhas and Raya Jalabi, "Ukraine Crisis: Why Russia Sees Crimea as Its Naval Stronghold," *The Guardian*, March 7, 2014; and Victor Zaborsky, *Crimea and the Black Sea Fleet in Russian-Ukrainian Relations*, Cambridge, Mass.: Harvard University, Kennedy School of Government, CSIA Discussion Paper 95-11, September 1995.

Presence of Russian Compatriots

Although the presence of Russian compatriots no doubt played a role in the war in the Donbas, their mere presence in the region was not a proximate driver to the escalation to conflict; rather, it was a structural contributor. As of March 2014, for instance, 38 percent of the population near Donetsk identified as ethnic Russian, and the majority of the area's population were Russian speakers.[49] The presence of Russian-speaking Ukrainians, ethnic Russians, and other compatriots in the Donbas was not a new phenomenon in early 2014. These populations had long coexisted in eastern Ukraine peacefully. The presence of Russian compatriots contributed to escalation only in the context of other related developments:[50] (1) the leadership change in Kyiv and the new government's pursuit of policies that might have triggered sensitivities in the Donbas, such as the repeal of a 2012 language law; (2) increased Ukrainian threat perceptions related to separatism in the wake of Crimea; and (3) Russian efforts to foment unrest in the Donbas. Relatedly, the Kremlin also probably sensed that it would incur reputational costs for failing to intervene on behalf of its compatriots in the Donbas whom Russian officials claimed were in danger.[51] The Kremlin likely felt pressure to fulfill its commitments to protect Russian compatriots abroad—promises that it had publicly affirmed through rhetoric and policies in the years before the Donbas conflict.[52]

---

[49] Piotr Zalewski, "Russian Separatism Gains Ground in Eastern Ukraine," *TIME*, March 19, 2014.

[50] Michael Kofman, Katya Migacheva, Brian Nichiporuk, Andrew Radin, Olesya Tkacheva, and Jenny Oberholtzer, *Lessons from Russia's Operations in Crimea and Eastern Ukraine*, Santa Monica, Calif.: RAND Corporation, RR-1498-A, 2017, p. 20.

[51] Steve Wilson, Peter Foster, and Katie Grant, "Ukraine as It Happened: Urgent Calls for Calm as West Faces Biggest Confrontation with Russia Since Cold War," *The Telegraph*, March 2, 2014.

[52] Igor Zevelev, *Russian National Identity and Foreign Policy*, Washington, D.C.: Center for Strategic and International Studies, Russia and Eurasia Program, December 2016, pp. 12–16.

Power Preponderance in Russia's Favor

In the case of Ukraine and Russia in the period prior to escalation in the Donbas, the power preponderance within this dyad favored Russia.

Ukraine's economy was a fraction of the size of Russia's in the year prior to the conflict when measuring in terms of gross domestic product (GDP). According to World Bank data, the Ukraine's GDP was $184.3 billion in 2013; Russia's neared $2.3 trillion.[53] As already noted, Ukraine's economy was highly dependent on Russia's—particularly in the energy sector, a lever that Russia exercised to influence Ukrainian policy (as we explore elsewhere in this appendix).[54] In the military sphere, Russia possessed approximately "four times as many soldiers . . . twice as many tanks . . . and more than six times as many combat aircraft" as Ukraine in early 2014.[55] Russia's defense budget was nearly 50 times that of Ukraine's. Furthermore, the equipment that Ukraine did have was aging, and its military was hindered by a need for institutional reforms.[56] Lastly, by spring 2014 (when the conflict first escalated), many of the modernization and professionalization reform efforts that the Russian military initiated in 2008 to address operational shortcomings that came to light from its conflict with Georgia had already materialized.[57]

The obvious Russian power preponderance over Ukraine is likely to have influenced the Kremlin's willingness to act from a position of strength in the relationship with Kyiv. Given that Moscow used lower energy prices and a $15 billion loan to Ukraine as an inducement to ditch the AA in the fall and winter 2013, the Kremlin was well aware of Ukraine's economic woes.[58] Likewise, the Kremlin's perceptions of Ukraine's military inferiority and political weaknesses were confirmed by the experience of the Crimean take-

---

[53] World Bank, "World Bank Open Data," webpage, undated-a.

[54] Menon and Rumer, 2015, pp. 41–43.

[55] Charles Recknagel, "A Side-by-Side Comparison of the Russian and Ukrainian Militaries," *The Atlantic*, March 19, 2014.

[56] Recknagel, 2014.

[57] Keir Giles, *Assessing Russia's Reorganized and Rearmed Military*, Washington, D.C.: Carnegie Endowment for International Peace, May 3, 2017.

[58] D'Anieri, 2019, p. 215.

over, during which Russian forces encountered little resistance.[59] However, it is unlikely that the disparity in relative strength in Russia's favor alone motivated the decision to escalate in eastern Ukraine; the power imbalance between Russia and Ukraine (or the power preponderance in Russia's favor) had existed for years without Russia taking similar actions. That said, it is conceivable that a Ukraine with more military and economic power might have acted as more of a deterrent against escalation.

Russian Threat Perceptions and Status Concerns

Russia Perceives Increased External Security Threats and Acute Uncertainty About the Future

The Kremlin had long warned both Kyiv and Brussels that it perceived closer Ukrainian integration with European political and military institutions as a threat. And yet, Russia never exercised hard-power instruments against Ukraine in response to developments in this sphere prior to 2014. Moscow's behavior in the year leading up to the Donbas conflict suggests that its expressed anxieties tied to Ukraine's signing of the AA were at least partly authentic. For instance, in the summer of 2013, when it appeared that Yanukovych was still proceeding with the AA, Russia levied so many major trade sanctions on Ukraine that a Ukrainian manufacturers trade group described them as a "complete halt on Ukrainian exports."[60] Furthermore, the Kremlin threatened increased tariffs on Ukrainian imports if Kyiv proceeded with the AA.[61] As noted above, in exchange for abandoning the EU deal, Moscow also offered Kyiv such significant financial inducements as a major reduction in gas prices and a $15 billion loan.[62]

The mass demonstrations and subsequent violence that erupted in Ukraine after its government announced its decision to suspend efforts to sign the AA introduced acute uncertainty into the Kremlin's calculus for

---

[59] David M. Herszenhorn, Patrick Reevell and Noah Sneider, "Russian Forces Take Over One of the Last Ukrainian Bases in Crimea," *New York Times*, March 22, 2014.

[60] Charap and Colton, 2017, p. 118.

[61] Charap and Colton, 2017, pp. 118–119.

[62] Menon and Rumer, 2015, p. 51.

Ukraine.[63] For Moscow, the massive antigovernment demonstrations in Kyiv might have touched on Kremlin insecurities about its own regime stability. Furthermore, the pro-Western (and, in some cases, anti-Russian) character of the protests in Ukraine also might have made Russian authorities uneasy. Existing evidence points in the direction that the Kremlin "pressured the Yanukovych government to forcibly repress the protests, but it is unclear how strong such pressure was."[64] Additionally, Russia deployed tens of thousands of military troops to its border with Ukraine in late February and early March 2014.[65] Such behavior serves as an indicator that the Kremlin was anxious about the unpredictability of the situation in Ukraine and experienced heightened threat perceptions introduced by the unrest in Ukraine but was not yet motivated to intervene.

It was only after the abrupt departure of Yanukovych in February 2014 and his replacement with a pro-Western government that was committed to signing the AA that the Kremlin decided to exercise instruments of hard power.[66] The ministers in Yanukovych's government overwhelmingly originated from Ukraine's southern and eastern regions, which are home to a larger proportion of Russian speakers and ethnic Russians and are often affiliated with a more eastern-leaning outlook. In contrast, a majority of the new government in Kyiv hailed from Ukraine's western regions;[67] some of these new officials held Ukrainian nationalist positions, and a small number had connections to Ukraine's far right. This likely caused significant pique in Moscow. Only two days after coming to power, the new parliament voted to overturn a 2012 law that allowed Ukrainian provinces to use other languages in formal settings.[68] The law ultimately remained in force, but the effort to repeal it might have confirmed Russian anxieties. As discussed

---

[63] "Ukraine Protests After Yanukovych EU Deal Rejection," *BBC News*, November 30, 2013; and Ann Taylor, "Days of Protest in Ukraine," *The Atlantic*, December 2, 2013.

[64] D'Anieri, 2019, p. 225.

[65] Simon Tisdall and Rory Carroll, "Russia Sets Terms for Ukraine Deal as 40,000 Troops Mass on Border," *The Guardian*, March 30, 2014.

[66] Götz, 2015, p. 5.

[67] Charap and Colton, 2017, p. 125.

[68] Tsygankov, 2015, p. 291.

already, the renewed possibility that Ukraine would sign the AA, move out of Russia's orbit, and someday join NATO was viewed by the Kremlin as a threat to Russian interests. Lastly, it was not only the acuteness of the perceived threats that drove Russia to use hard power in eastern Ukraine, but also the fact that Moscow had exhausted other soft-power instruments, such as economic threats.[69]

Reputational Costs for Russia for Nonintervention and Russian Dissatisfaction with Its Place in the International System

Russia's decision to escalate to conflict in the Donbas also might have been partially motivated by reputational costs that Moscow believed it would incur if it did not intervene using military means. There were regional and international dimensions to these costs. The first relates to regional audiences situated in Russia's perceived sphere of interests, or the post-Soviet space where Russian policymakers have claimed that Russia has a unique right to exert authority.[70] In the years prior to the Donbas conflict, the Kremlin had very publicly committed to protecting Russian speakers, ethnic Russians, and other Russian compatriots. Russian officials reaffirmed these commitments in the context of the Ukraine crisis. For instance, Putin press secretary Dmitry Peskov stated that Russia could not sit idly as its compatriots were threatened, particularly in a neighboring country, and even went so far as to name Putin as "the main guarantor of security in the Russian world" in early March 2014.[71] In these circumstances, it is possible that the Kremlin felt its reputation would suffer a significant blow if Moscow did not deliver on its promises to protect compatriots in Ukraine, particularly in the eyes of regional audiences where Russian compatriots were most prevalent.

It is also possible that Putin could have been concerned about tarnishing his personal reputation in the eyes of regional audiences for failing to protect compatriots.

---

[69] Menon and Rumer, 2015, p. 83.

[70] Dmitri Trenin, "Russia's Spheres of *Interest*, Not *Influence*," *Washington Quarterly*, Vol. 32, No. 4, October 2009.

[71] "Peskov: Rossiya ne mozhet byt' v storone, kogda russkim grozyat nasiliem," *RIA Novosti*, March 7, 2014.

The Kremlin might have perceived that Russia would also bear significant reputational costs internationally if it did not intervene in the Donbas. Russia's self-identity as a great power is tied in part to its belief that it has unique authority over issues in its regional sphere of interest, particularly in neighboring Ukraine.[72] The West largely has not ascribed to this view. Rather, the West's position has been that Ukraine and other states in the post-Soviet space should be free to design their own destiny.[73] Russian decisionmakers have argued this position is hypocritical—that the West critiques Russian activities in Moscow's own backyard but promotes democratic movements globally in an effort to expand its own influence in the international system.[74]

This contrast in perspectives can be seen in the responses to the power transition in Kyiv in February 2014. After Yanukovych's departure, Russia declared that the interim government that came to power as a result of popular uprising was illegitimate; the West expressed approval of these events.[75] This suggests that it is possible that Russian decisionmakers viewed the unrest in Ukraine as a test of wills. Viewed through this prism, inaction in Ukraine could have undermined Russia's position as a regional hegemon.

Russia's dissatisfaction with its place in the international system was also clearly a factor in its decision to pursue the war in the Donbas. On March 18, Putin delivered a speech on Crimea's "reunification" with Russia, which echoed some of the broader themes he raised in the 2007 Munich speech. Putin portrayed the West's policy toward Russia as a new iteration of the Cold War–era containment policy. "We have every reason to assume that [policy], which was conducted in the 18th, 19th and 20th centuries, continues today," Putin bristled. "They are constantly trying to sweep us into a corner because we have an independent position, because we maintain it and because we call things like they are and do not engage in hypocrisy."

---

[72] D'Anieri, 2019, pp. 3, 10, 16.

[73] See, for instance, Joseph Biden, "Remarks by Vice President Biden in Ukraine," speech given in Kyiv, Ukraine, July 22, 2009.

[74] See, for instance, Sergei Lavrov, "Russia's Foreign Policy Philosophy," *International Affairs*, March 2013.

[75] Charap and Colton, 2017, pp. 125–126.

Yet, with Ukraine, the West "crossed the line . . . acting irresponsibly and unprofessionally." Ultimately, "Russia found itself in a position it could not retreat from. If you compress the spring all the way to its limit, it will snap back hard."[76]

Domestic Drivers

Opposing State Experiences Domestic Political Instability

Ukraine was experiencing significant domestic political instability at the time of the escalation. Broadly speaking, by the time the conflict in the Donbas commenced in the spring 2014, Ukraine had experienced mass demonstrations, violence as a result of the protests, the abandonment of its executive, a transition of power in Kyiv, and the annexation of a part of its sovereign territory. Although this broader instability likely contributed to the escalation of conflict between Russia and Ukraine, the separatist activities that took place in eastern Ukraine in spring 2014 were likely more-proximate drivers of both Ukraine's and Russia's decisions to escalate. From the Ukrainian perspective, political instability in the east (much of which was at the hands of separatists calling for independence from Kyiv) represented a threat to the country's territory. The Ukrainian government's response to this instability in the form of the ATO directly contributed to escalation. The Kremlin, on the other hand, viewed the governmental turmoil precipitated by Yanukovych's departure as an opportune time to intensify support to (or deployment of) separatist forces in the Donbas, thereby also contributing to escalation.[77]

Conclusion

Our analysis indicates that structural factors such as a shared land border, the status of Ukraine as a former Soviet republic, the presence of Russian compatriots in the Donbas region, the salience of the Donbas to Russia, a

---

[76] President of Russia, 2014a.

[77] D'Anieri, 2019, pp. 227–228.

power preponderance in Russia's favor, and Russia's dissatisfaction with its place in the international system were important drivers in escalation to conflict between Ukraine and Russia in 2014, contributing to long-term discord between Kyiv and Moscow. However, our analysis also suggests that these factors, although necessary, were likely insufficient on their own to motivate Russia to escalate to conflict. They had existed long before 2014, and although they resulted in soured and tense relations between Kyiv and Moscow, the two parties had not previously engaged in conflict. This suggests that other, more-immediate or proximate factors ultimately drove Russia and Ukraine to hostilities. Specifically, Russian perceptions of increased external security threats, the Kremlin's uncertainty about the future, domestic political instability in Ukraine, the threats to the Donbas representing a critical territorial issue to Ukraine, and Russian concerns with incurring reputational costs for nonintervention likely served as the match to the powder keg.

Militarized Conflict Case Study: Russia-Georgia War, 2008

The short war between Russia and Georgia in August 2008 was, at the time, the largest outbreak of fighting in Europe since the Kosovo War in 1999. The conflict was initially centered in South Ossetia but spread to Abkhazia as well. Both regions were de facto states that were part of internationally recognized Georgian territory, but they declared independence and gained autonomy from Tbilisi in the early 1990s as a result of intergroup violence that spiraled into civil conflicts.[1]

Tensions between Georgia and South Ossetian militia forces, with the latter backed by the Russian Federation, had been simmering for several months prior to the escalation that led to all-out war in August 2008. This period was characterized by sporadic interethnic violence and artillery fire exchanges in South Ossetia, Russian violations of Georgian air space, and both Russian and Georgian military exercises. Russia-Georgia relations had been increasingly tense since the spring, when Russia (following Kosovo's contested independence and the NATO Bucharest Summit) took steps to increase its political and military control over the two de facto states. The circumstances of the outbreak of the August 2008 war remain contested. For Russia, the war was a defensive response based on its self-proclaimed responsibility to protect vulnerable minority groups (most of

---

[1] A *de facto state* is defined as a region that has received internal popular support and has achieved the capacity to provide governmental services to its population within a defined territorial area over which it has control but lacks international recognition of its independence. See Scott Pegg, *International Society and the De Facto State*, Brookfield, Vt.: Ashgate, 1998.

whom were also Russian compatriots) from alleged genocide. For Georgia, it was a planned Russian invasion of sovereign Georgian territory and, more broadly, an assault on the West. For the South Ossetians, the war was a renewed attempt of oppression by the Georgian state.[2]

The war began on August 7, 2008, with a Georgian offensive on the regional capital Tskhinvali and the surrounding villages in South Ossetia, followed by the initial Russian response.[3] Afterward, a Russian-South Ossetian counterattack gained momentum as military and infrastructure targets throughout Georgia were attacked. The Russian ground operation was supported by air and naval assaults and augmented by a cyber offensive against Georgia's government and financial system.[4] Simultaneously, a Russian ground invasion of Tbilisi-administered territory adjacent to Abkhazia pushed deep into uncontested Georgian territory.[5] A ceasefire agreement brokered by French President Nicolas Sarkozy brought an end to the fighting.[6] Russian forces finally completed their withdrawal from occupying positions in Tbilisi-administered territory on October 8, 2008.

---

[2] Gerard Toal, *Near Abroad: Putin, the West, and the Contest over Ukraine and the Caucasus*, New York: Oxford University Press, 2017, p. 127.

[3] An independent, fact-finding report commissioned by the EU concluded that the Georgian government initiated open hostilities on August 7, 2008. But it also concluded that the Russian government was partly responsible for the conditions that led to the beginning of the large-scale conflict in Georgia (Independent International Fact-Finding Mission on the Conflict in Georgia, *Independent International Fact-Finding Mission on the Conflict in Georgia Report*, Vol. I, Brussels: Council of the European Union, September 2009).

[4] Ronald D. Asmus, *A Little War That Shook the World*, New York: Palgrave Macmillan, 2010, pp. 166–167.

[5] It is conventional to use the term *Tbilisi-administered territory* to refer to areas of Georgia outside the breakaway regions of Abkhazia and South Ossetia.

[6] President Sarkozy was acting in his capacity as president of the Council of the European Union.

Drivers

The August 2008 war was a complex and multidimensional event caused by nine key drivers:

1. a shared land border between Georgia and Russia
2. Georgia's location in geographic proximity to the Russian Federation and its status as a former Soviet republic
3. a conflict involving a territorial issue for Georgia
4. Russia's dissatisfaction with its role in the international system
5. perceived reputational costs for Russia for nonintervention
6. Russia's perceived increased external security threats
7. Russia's acute uncertainty about the future
8. the presence of a Russian military installation
9. the presence of Russian compatriots in South Ossetia.

Georgia's sharing of a land border with Russia and its status as a former Soviet republic resulted in a legacy of Georgia being a contested territorial space in which disparate ethnic groups competed for power and security. These factors also represented underlying structural factors for escalation. The collapse of the Soviet Union set the conditions for the territorial fragmentation of Georgia in the early 1990s. Both Abkhazia and South Ossetia emerged as de facto states that depended on Russian patronage to ensure their continued existence. Independent Georgia never enjoyed full control over its internationally recognized territory. Under President Saakashvili, Georgia sought to recover Abkhazia and South Ossetia, which led to escalated levels of violence and polarization, and punitive Russian responses. By 2007, Russian President Putin was loudly voicing dissatisfaction with Russia's place in the international system and setting out on a course of foreign policy assertiveness, representing a trigger for escalation in this case. Other immediate triggers for escalation resulted from a series of decisions made in early 2008 by the United States and its European allies that added to Russia's already existing resentment and insecurity. The prospect of Georgia's NATO membership demonstrated Saakashvili's success in breaking free from the legacy of dependence on Russia, which intensified fears in the Kremlin. The long-standing presence of Russian peacekeepers in Abkhazia and South Ossetia allowed the de facto authorities to establish government structures and allowed Russia to carry out military buildups in the regions. After Georgian forces initiated the offensive in South Ossetia on August 7, 2008, the threat to Russian peacekeepers compelled Russia to counterattack. Finally, Russia's self-designated responsibility to protect its compatriots provided an immediate pretext for the invasion on August 8.

Geographic and Territorial Drivers

Geographic Drivers: Shared Land Border and Former Soviet Republic Status

Georgia's status as a former Soviet republic and its sharing of a land border with Russia were two structural drivers of the August 2008 war. The legacy of the Soviet Union as an imperial complex that created the borders of the current Georgian state (and the distinct ethnoterritorial units within it) had a direct and long-term impact on the escalatory dynamics that unfolded in the summer of 2008. The wars over South Ossetia in 1991–1992 and Abkhazia in 1992–1993 were characterized by ethnic cleansing, population displacement, and destruction of infrastructure. The secessionist forces drew from preexisting patron-client relationships with the Soviet establishment to either win the wars or bring them to stalemates.[7] Russia brokered the ceasefire agreements to both conflicts. The two separate sets of monitoring and peacekeeping arrangements institutionalized an enduring presence for Russian peacekeeping forces on the ground in Abkhazia and South Ossetia. Georgia, at a time of extreme weakness, accepted the arrangements to bring an end to the wars.[8]

---

[7] Soviet (and subsequently Russian) forces supplied both Georgian and secessionist forces during the wars through both opportunistic individuals and entire units. This fact should be understood within the context of a collapsing Soviet Union, in which central control over the Soviet military weakened significantly. Much of the military support that the Abkhaz and Ossetians received was probably freelance, coming from unemployed Russian officers (Thomas de Waal, *The Caucasus: An Introduction*, New York: Oxford University Press, 2010, p. 160; and Stuart J. Kaufman, "Georgia and the Fears of Majorities," in *Modern Hatreds: The Symbolic Politics of Ethnic War*, Ithaca, N.Y.: Cornell University Press, 2001, p. 90).

[8] The two separate sets of monitoring and peacekeeping arrangements with the UN and Organization for Security and Co-operation in Europe (OSCE), respectively, were flawed and limited instruments: They had local composition, limited international

Thereafter, the two de facto states considered the conflicts with the Georgian state effectively resolved; however, the breakaway regions lacked international recognition of their independence and therefore the external legitimacy of full statehood.

In addition, Georgia directly borders the Russian North Caucasus, which has been the most volatile area of the country since the Soviet collapse. The brief war between ethnic Ingush and Ossetians in 1992 over a disputed area around Vladikavkaz called Prigorodnyy was the first ethnic conflict on Russian Federation territory after the Soviet breakup. The Russian government has also fought two wars against separatists in Chechnya. The first war (1994–1996) resulted in an embarrassing defeat for Russia and the republic's de facto independence. In the second war (1999–2000), the Russian army used overwhelming force against Chechnya, which resulted in an enormous loss of civilian lives, population displacement, and destruction of infrastructure. Although the large-scale conflict ended, there still has been low-level fighting against the remaining rebels.

Thus, the prospect for a future Georgian government to try to reclaim Abkhazia and South Ossetia by using force—and the attendant potential spillover of violence and population displacement into the North Caucasus that it could cause—likely contributed to Russia's sense of insecurity regarding Georgia's future.

Territorial Drivers: Territorial Issue for Georgia

The conflict involved direct implications for Georgia's territorial integrity, and President Saakashvili's concerted effort to establish full control over Georgia's internationally recognized territory was a key driver of the August 2008 war. After the breakup of the Soviet Union and the civil wars of the early 1990s that led to the birth of the two de facto states of Abkhazia and South Ossetia, the Georgian government never enjoyed control over the two regions. Georgian forces controlled only a small sliver of the mountainous Kodori Gorge in Abkhazia and a patchwork area of less than half of South Ossetia.[9] The lack

---

participation, and questionable neutrality. Still, the United States and several powerful European states sanctioned Russia to be the main peacekeeping force on the ground in Abkhazia and South Ossetia (Asmus, 2010, p. 65).

[9] Asmus, 2010, p. 30.

of Georgian control and the presence of Russian peacekeepers enabled the de facto authorities to assert their own local cultures and establish administrative structures.[10] Both Abkhazia and South Ossetia also shared a border with the Russian Federation, which facilitated deeper cultural, economic, political, and security ties with ethnic kin in neighboring Russian regions, such as North Ossetia.[11] The two breakaway regions developed into Russian client statelets that relied heavily on external patronage to ensure their survival.

Georgian politics was transformed by the Rose Revolution of November 2003, which brought Saakashvili to power. As president, he pursued two main policy goals: Recover the territories not controlled by the Georgian government and seek membership for Georgia in the EU and NATO.[12] Saakashvili did not believe he could survive as president if he failed to defend Georgian citizens in the areas that the government still controlled or if he lost the breakaway territories permanently.[13] To that end, his government took a series of bold actions against the de facto states in an effort to reassert Georgia's sovereignty. These moves created the conditions for renewed polarization and led to an increase in interethnic violence in both de facto states, especially South Ossetia.[14] Still, the territorial disputes remained unresolved. Before the August 2008 war, the official Russian position was that both Abkhazia and South Ossetia were unsettled internal territorial conflicts within the state of Georgia. Yet, the realities on the ground suggested Moscow favored the two de facto regimes.[15]

---

[10] Pål Kolstø, "The Sustainability and Future of Unrecognized Quasi-States," *Journal of Peace Research*, Vol. 43, No. 6, 2006.

[11] During the Soviet era, many Ossetians living in Georgia became culturally Georgian as a means of social mobility. Most also held strong ties with ethnic kin in North Ossetia (Toal, 2017, pp. 35–36).

[12] Gearóid Ó Tuathail, "Russia's Kosovo: A Critical Geopolitics of the August 2008 War over South Ossetia," *Eurasian Geography and Economics*, Vol. 49, No. 6, 2008.

[13] Asmus, 2010, p. 29.

[14] Tuathail, 2008.

[15] The Georgian government never viewed the Russian peacekeepers in South Ossetia as impartial; from the mid-2000s, Georgian officials believed these forces were shielding Ossetian militias involved in attacks on Georgian villages (Roy Allison, "Russia Resurgent? Moscow's Campaign to 'Coerce Georgia to Peace,'" *International Affairs*, Vol. 84, No. 6, November 2008).

Russian Threat Perceptions and Status Concerns

Russian Dissatisfaction with Its Place in the International System and Reputational Costs for Russia for Nonintervention

The August 2008 war followed a period of rising tensions between Russia and the Euro-Atlantic community. By 2008, Russia was convinced that its national security interests were not being respected by Western powers, especially the United States, and that nothing it was doing diplomatically could stop the process of Western encroachment into its neighborhood. These factors set the conditions for escalation. Relations had deteriorated as a result of many events over the previous year: the U.S. announcement in April 2007 to develop missile defense systems in Eastern Europe, Russia's suspension of the Conventional Forces in Europe Treaty in December 2007, the recognition of Kosovo's independence from Serbia by the United States and several European powers in February 2008, and NATO's Bucharest Summit declaration in April 2008 regarding Georgia's and Ukraine's eventual membership in the Alliance. Thus, the pragmatic security cooperation that characterized Russian foreign policy in the early 2000s era after the September 11 terrorist attacks in the United States—which included a joint U.S.-Russian-Georgian counterterrorism operation in Georgia's Pankisi Gorge—gave way to a new agenda of assertiveness.[16] This agenda called for restoring Russia's great power status through exploiting its oil and gas wealth; independence from U.S.-dominated institutions; maintaining preeminent influence in the post-Soviet space, including in the southern Caucasus; and promoting a multipolar world order.[17] The essence of Russia's position on its place in the post–Cold War security order was clearest in President Putin's February 2007 speech in Munich.[18] According to Putin, U.S. unilateralism in Iraq and unipolarity in world affairs threatened

---

[16] The operation against Chechen fighters and suspected Islamic extremists in the spring of 2002 represented the high-water mark of cooperation in U.S.-Russian-Georgian relations (Toal, 2017, p. 110). Regarding the period of cooperation between Russia and the West in the early 2000s, see Tsygankov, 2015.

[17] Jeffery Mankoff, *Russian Foreign Policy: The Return of Great Power Politics*, Lanham, Md.: Rowman & Littlefield, 2009.

[18] William H. Hill, *No Place for Russia: European Security Institutions Since 1989*, New York: Columbia University Press, 2018, pp. 251–252.

international security, the OSCE had been transformed into an instrument designed to advance only Western interests, and NATO's eastward enlargement was a serious provocation that created new dividing lines in Europe.[19]

If Georgia were successful in the conflict, Russia risked losing its clout as regional hegemon. If a post-Soviet state that was orders of magnitude smaller than Russia succeeded militarily, Russia could no longer claim great power status even in its near abroad, let alone globally. In an interview conducted shortly after the conflict, then-President Dmitry Medvedev asserted, as a pillar of his foreign policy strategy, that Russia would pursue its "privileged interests" in key regions.[20] Moreover, local conditions added pressure to Moscow's decisionmaking. Given the presence of Russian citizens in South Ossetia, conflict there raised the risk of potential Russian deaths. The inability to protect Russian nationals almost certainly would have been a severe blow to Russia's credibility. Medvedev highlighted this issue in the same postconflict interview, claiming that "protecting the lives and dignity of our citizens, wherever they might be, is an unquestionable priority for our country."[21]

Russia Perceives Increased External Security Threats and Acute Uncertainty About the Future

In the run-up to the August 2008 war, there were two events that inflamed already existing tensions in Russia-West relations. The first was Kosovo's contested declaration of independence in February 2008, which ended a period characterized by a lack of success of the EU, OSCE, NATO, and UN mechanisms seeking an agreement between Serbia and its breakaway province. Russia adamantly opposed Kosovo's independence, calling it a violation of Serbia's territorial integrity.[22] Russian leaders argued that the move might serve as precedent for Chechnya or other potential separatist entities in the Russian Federation, could create a spillover to other territories in the Balkans (such as Republika Srpska), and could buttress the claims for rec-

---

[19] President of Russia, 2007. The occasion of Putin's attendance marked the first time a Russian president had been invited to the annual Munich Security Conference.

[20] President of Russia, "Interview Given by Dmitry Medvedev to Television Channels Channel One, Rossia, NTV," webpage, August 31, 2008b.

[21] President of Russia, 2008b.

[22] Hill, 2018, p. 257.

ognition of the de facto states in Georgia and elsewhere in the post-Soviet space.[23] Facing the possibility of a Russian veto in the UN Security Council, the United States and the EU removed the question of Kosovo's future status from the UN framework.[24] Kosovo declared its independence on February 18 and was immediately recognized by the United States and most EU states. Western leaders argued that Kosovo represented a unique case and that recognition should be awarded because it had built effective democratic structures.[25] In response, the Russian State Duma passed a motion urging the Kremlin to recognize Abkhazia and South Ossetia as independent states. On February 22, 2008, Putin reportedly told Saakashvili that Russia would be forced to answer the West on Kosovo and that Georgia would be part of that response.[26]

The independence of Kosovo contributed to the escalatory dynamic leading to the August 2008 war insofar as it gave Russia an opportunity to expose what it considered a clear instance of a Western double standard; the Euro-Atlantic community supported Kosovo's independence from Serbia, a traditional Russian ally, but not Abkhazia's and South Ossetia's independence from Georgia, which had sought to join Western institutions and had developed close bilateral relations with the United States.[27] The occasion of the war enabled the Kremlin to follow through on its promise to respond to the example of Kosovo.

The April 2008 NATO Bucharest Summit declaration that Georgia and Ukraine "will become" members of NATO was the second event that hastened the escalation spiral. The declaration was unprecedented in many

---

[23] In January 2006, Putin announced that there was a need for "universal principles" to settle the unresolved conflicts ranging from Kosovo to Abkhazia and South Ossetia (President of Russia, "Press Conference for the Russian and Foreign Media," speech, Moscow, January 31, 2006).

[24] Hill, 2018, p. 257.

[25] Nina Caspersen and Gareth Stansfield, eds., *Unrecognized States in the International System*, New York: Routledge, Exeter Studies in Ethno Politics, 2011, p. 61.

[26] According to the Georgian record, Putin explained to Saakashvili: "You know we have to answer the West on Kosovo. And we are very sorry, but you are going to be part of that response. Your geography is what it is" (Toal, 2017, p. 155).

[27] Tuathail, 2008, p. 683.

ways.[28] First, it signaled NATO's intent to expand into territories that were entangled with Russian and Soviet history, identity, and territory. No country that was part of the original Soviet Union had ever joined "the West," as signified by admission into NATO.[29] Second, NATO considered incorporating Georgia despite its long-standing internal ethnoterritorial divisions and polarization, which were likely going to be exacerbated by the move.[30] Finally, NATO pursued Georgia's membership despite consistent warnings from Moscow that NATO's enlargement there would cross a "red line" and prompt a dramatic shift in Russian policy toward the Alliance.[31] At the heart of Russia's hostility toward Georgia's NATO membership is a long-standing fear of its encirclement by conspiring external forces; with Georgia incorporated under NATO's security umbrella, the country would be more likely to serve as a platform for the deployment of U.S. forces and weapon systems potentially targeted at Russia.

Russia's rising tensions with the Euro-Atlantic community in 2007–2008 mirrored the deterioration in Russian relations with Georgia. The source of the discord was Saakashvili's attempt to reunify the country and seek NATO membership. He vigorously pursued a reform agenda that included modernizing the military to bring it up to NATO standards. In this effort, Georgia received substantial aid from Western countries, especially the

---

[28] The Bucharest Summit of April 2008 was the first time a Russian president had been invited to a NATO summit. At his press conference after the summit, Putin reiterated Russia's position on enlargement to Georgia and Ukraine: "We view the appearance of a powerful military bloc on our borders . . . as a direct threat to the security of our country. The claim that this process is not directed against Russia will not suffice. National security is not based on promises" (President of Russia, "Press Statement and Answers to Journalists' Questions Following a Meeting of the Russia-NATO Council," transcript, April 4, 2008a; and Hill, 2018, p. 262).

[29] In 1922, an independent Georgia was forcibly annexed by Soviet Russia and later incorporated into the Soviet Union as an original member (Toal, 2017, p. 7).

[30] NATO's own post–Cold War enlargement study stipulates that the unresolved territorial disputes of aspirant members would be a factor in determining their suitability for membership (NATO, "Study on NATO Enlargement," webpage, September 3, 1995).

[31] Russian Foreign Minister Sergei Lavrov asserted that the possible entry of Georgia and Ukraine would bring about a tremendous "geopolitical shift" requiring Russia to "revise its policy" (Sergei Lavrov, "NATO Expansion a Huge Mistake," *Interfax*, December 12, 2006).

United States. Saakashvili's government figured prominently in the Bush administration's "freedom agenda" of promoting democratization across the globe.

Georgia's efforts to regain control provoked further radicalization in the de facto states and punitive measures by Russia. In 2006, Russia imposed an embargo on crucial Georgian exports, including wine, and blocked transport and postal links.[32] Ethnic Georgians living in Russia were subjected to harassment and deportation.[33] In March 2008, Russia ended its unilateral sanction agreement against Abkhazia that dated back to 1996.[34] On all sides, the period prior to the August 2008 war involved aggressive actions leading to escalating competition, polarized populations, and simmering levels of violence that eventually boiled over in late summer.

Nature of the Relationship Between Russia and the Opposing State

Presence of Russian Military Installation

By 2008, Russian peacekeeping forces had already been deployed to Abkhazia and South Ossetia for a decade and a half. After Georgian forces initiated the offensive against Tskhinvali and the surrounding South Ossetian–controlled villages on August 7, 2008, the Kremlin claimed its response was driven by self-defense of the existing Russian peacekeepers on the ground.[35] Although the peacekeepers had been in place for many years, the immediate threat to their lives provided a short-term spark to the escalatory dynamic.

The Russian troop presence was authorized by the monitoring and peacekeeping mechanisms implemented after the civil wars in Georgia in the early 1990s. Russia was permitted 3,000 troops in Abkhazia, with no Georgian participation of any kind. In South Ossetia, a joint peacekeeping

---

[32] Asmus, 2010, p. 73.

[33] Human Rights Watch, *Singled Out: Russia's Detention and Expulsion of Georgians*, Vol. 19, No. 5(D), New York, October 2007.

[34] Asmus, 2010, p. 108.

[35] Allison, 2008.

force was formed with a battalion of Russian, South Ossetian, and Georgian forces under Russian command.

Presence of Russian Compatriots

The commitment to defend Russian compatriots and citizens living abroad (i.e., Russian passport holders in South Ossetia) against an immediate threat was a crucial driver of the August 2008 war. Even before the outbreak of war, the official Russian position that both Abkhazia and South Ossetia were unresolved conflicts within the internationally recognized territory of Georgia was complicated by a change to Russian citizenship law in May 2002, which streamlined the procedures for residents in the de facto states to obtain Russian nationality.[36] Under the new law, people were able to apply for passports from their home localities and without establishing residency in the Russian Federation. The change resulted in a large increase in the number of Russian passport holders in Abkhazia and South Ossetia.[37] Nearly the entire populations of both regions had acquired Russian citizenship by August 2008.[38] The policy change proved significant because thereafter a Russian leader could claim a right to intervene in Russia's neighborhood to protect Russian citizens.[39] In mid-July 2008, new President Dmitry Medvedev released Russia's Foreign Policy Concept, which emphasized the importance of the protection of the rights and interests of Russian citizens in other countries.[40]

Russia framed its initial military response to the Georgian offensive against South Ossetia on August 7, 2008, as a "rescue mission," which was justified under Article 51 of the UN Charter in self-defense.[41] Medvedev proclaimed that Georgia's actions constituted an act of aggression against

---

[36] Toal, 2017, p. 139.

[37] Vincent Artman, "Documenting Territory: Passportisation, Territory, and Exception in Abkhazia and South Ossetia," *Geopolitics*, Vol. 18, No. 3, 2013.

[38] Independent International Fact-Finding Mission on the Conflict in Georgia, 2009.

[39] Toal, 2017, pp. 140–141.

[40] Ministry of Foreign Affairs of the Russian Federation, *The Foreign Policy Concept of the Russian Federation*, translation, Moscow, July 12, 2008.

[41] Toal, 2017, pp. 12–13.

Russian peacekeepers and the civilian population in South Ossetia. By then, through a naturalization process that had been taking place over many years, over 90 percent of the residents of South Ossetia were Russian citizens.[42] Medvedev asserted that, as president of the Russian Federation, he was duty bound to protect the lives of Russian citizens wherever they might be.[43] As events unfolded in the following days, Russia's official narrative shifted from self-defense to the claim that Georgian forces were committing "genocide" against South Ossetians.[44] Subsequently, Russia responded to growing international criticism of its occupation of large parts of Georgian territory by framing its actions within the "Responsibility to Protect" doctrine, even though it opposed NATO evoking these norms in the 1999 Kosovo intervention.[45]

Conclusion

The August 2008 war followed a period of long-simmering tensions between Russia and Georgia that escalated quickly when fighting broke out between Georgian forces and South Ossetian forces after many years of polarization, radicalization, and intermittent violence in the separatist region. Russia responded with overwhelming force by mounting a large-scale invasion that devastated Georgia. The war resulted in a military victory for Russia and enabled Moscow to consolidate control over Abkhazia and South Ossetia. However, Russia's actions damaged its international reputation and alarmed its neighbors and partners. On August 26, 2008, Russia recognized Abkhazia and South Ossetia as independent states, a move that has been followed by few other countries.[46]

---

[42] Artman, 2013, p. 684.

[43] Toal, 2017, p. 181.

[44] The EU's report on the war dismisses Russian allegations of genocide against South Ossetians as unfounded and unsubstantiated (Independent International Fact-Finding Mission on the Conflict in Georgia, 2009).

[45] Toal, 2017, p. 185.

[46] "Russia Recognizes Abkhazia, South Ossetia," *Radio Free Europe/Radio Liberty Newsline*, August 26, 2008. Russia's recognition of Abkhazia and South Ossetia has been

The war was caused by a variety of drivers, from long-standing structural conditions to near-term or immediate triggers or sparks. The war would have been unlikely without the previous territorial fragmentation of Georgia and the emergence of pro-Russian client statelets, which resulted from Georgia's status as a former Soviet republic and its shared border with Russia. Also contributing to the escalation were Russia's growing sense of resentment regarding its place in the international system and its acute sense of insecurity and uncertainty about the future resulting from Kosovo's contested independence and NATO's anticipated eastward enlargement to Georgia and Ukraine at some point in the future. After Georgian forces initiated the offensive on Tskhinvali on August 7, 2008—and with the Kremlin fearing that it would incur reputational costs for nonintervention—Russian leaders felt compelled to uphold their previous commitments to respond to threats against Russian peacekeepers and Russian compatriots and citizens in South Ossetia; their presence in the conflict zone and the need to avoid reputational costs for nonintervention constituted the immediate drivers of escalation to war.

Under Saakashvili, Georgia made a concerted effort to recover Abkhazia and South Ossetia, which represented important territorial issues for Georgia, and to seek membership in Western institutions, specifically NATO and the EU. His aspiration to geopolitically reorient Georgia away from Russia provoked a strong reaction both within the two de facto states and within the Kremlin. Moscow was willing to exploit the unresolved territorial conflicts to undercut Saakashvili and prevent what Russia understood as further Western encroachment on its borders. Russia's actions sent a clear message to other states desiring to integrate with the West: Doing so could lead to their territorial dismemberment.

---

followed by only a handful of pro-Russia countries: Nauru, Nicaragua, Syria, Tuvalu, Vanuatu, and Venezuela. Tuvalu subsequently withdrew its recognition in March 2014. See "Tuvalu Retracts Abkhazia, S. Ossetia Recognition," *Civil Georgia*, March 31, 2014.

Militarized Conflict Case Study: Tajik Civil War, 1992–1997

In September 1991, the Tajik SSR declared its independence from the Soviet Union.[1] In the following nine months, internal tensions between a secular government of former Communist Party members[2] and opposition forces bringing together nationalist, Islamist, and prodemocracy parties mounted gradually,[3] leading to the outbreak of civil war in May 1992.[4] The nationalist forces associated with the "Rastakhiz (Rebirth) movement called for a revival of a Tajik identity, culture and language," the Islamic Renaissance Party "advocated the Islamisation of Tajik society and politics," while the Democratic Party advanced an agenda focused on democratizing the country.[5]

Throughout the conflict, both government and opposition forces relied on foreign sponsors, adding another layer of complexity to the already con-

---

[1] Bobi Pirseyedi, "The Conflict in Tajikistan," in *The Small Arms Problem in Central Asia: Features and Implications*, Geneva: United Nations Institute for Disarmament Research, 2000, p. 39.

[2] During Perestroika, Tajikistan "gained a reputation as having one of the most conservative and tenacious party elites in the Soviet Union. A local communist elite dependent for its survival on subsidies from Moscow was hardly likely to articulate nationalist or separatist aspirations" (Barnett R. Rubin, "The Fragmentation of Tajikistan," *Survival*, Vol. 35, No. 4, 1993, p. 76).

[3] Pirseyedi, 2000, p. 40; Rubin, 1993, p. 76.

[4] In the context of the civil war, the Uzbek minority and the Russian speakers in Tajikistan supported the Communist government in power (Rubin, 1993, p. 76).

[5] Rubin, 1993, p. 76. Also see Bess A. Brown, "The Civil War in Tajikistan, 1992–1993," in Mohammad-Reza Djalili, Frédéric Grare, and Shirin Akiner, eds., *Tajikistan: The Trials of Independence*, 1st ed., New York: Routledge, 1998; and Pirseyedi, 2000, p. 39.

voluted situation on the ground. The progovernment forces received external support from Russia and the other Central Asian governments that were aiming to restore stability in Tajikistan and maintain the status quo. In turn, the opposition forces received support from Iran, the Afghan mujahedin,[6] field commanders in Northern Afghanistan, and Islamic militants and sponsors from Pakistan and Saudi Arabia who were interested in advancing an Islamist agenda in Central Asia. Iran also intervened as a mediator in the conflict.[7]

As a result of the violence, between 20,000 and 40,000 people lost their lives,[8] some 600,000 were internally displaced, and approximately 100,000 became refugees.[9] The war officially came to an end in June 1997 when a peace accord was signed.[10] The most violent phase of the conflict took place between the spring of 1992 and the beginning of 1993,[11] when "the former Communist elite emerged as a winner,"[12] and the old-style Communist type of government was reinstated.[13] After this, with opposition fighters taking refuge in Afghanistan and conducting incursions across the border into

---

[6] Susan Clark, *The Central Asian States: Defining Security Priorities and Developing Military Forces*, Alexandria, Va.: Institute for Defense Analyses, IDA Paper P-2886, September 1993, p. 13.

[7] Shirin Akiner and Catherine Barnes, "The Tajik Civil War: Causes and Dynamics," in Abdullaev and Barnes, 2001, pp. 20–21.

[8] Dov Lynch, "The Tajik Civil War and Peace Process," *Civil Wars*, Vol. 4, No. 4, 2001, pp. 49–50. Clark (1993, p. 12) placed the number of victims as ranging between 20,000 and 70,000; Pirseyedi and Brown mention between 20,000 and 100,000 individuals killed (Pirseyedi, 2000, p. 42; and Brown, 1998).

[9] Lynch, 2001, pp. 49–50.

[10] Political Economy Research Institute, *Modern Conflicts: Conflict Profile, Tajikistan (1992–1998)*, Amherst, Mass.: University of Massachusetts Amherst, undated. For more details on the peace accord, please see Kamoludin Abdullaev and Catherine Barnes, eds., *Accord*: Vol. 10, *Politics of Compromise: The Tajikistan Peace Process*, London: Conciliation Resources, 2001.

[11] Brown, 1998.

[12] Pirseyedi, 2000, p. 42.

[13] Stéphane A. Dudoignon, "Political Parties and Forces in Tajikistan, 1989–1993," in Mohammad-Reza Djalili, Frédéric Grare, and Shirin Akiner, eds., *Tajikistan: The Trials of Independence*, New York: Routledge, 1998.

Tajikistan,[14] the fighting became concentrated on the Tajik-Afghan border. Low-intensity violent clashes between the government and the opposition forces continued until 1997 under the guise of guerrilla warfare.

When the civil war broke out in May 1992, Russian troops were already stationed on the Tajik-Afghan border and on the outskirts of Dushanbe, the capital, where the 201st Motorized Rifle Division (MRD) was based. These stationed troops were Soviet border and army units, respectively, that still answered to Moscow after the Soviet Union's collapse. The 201st, a standard Soviet mechanized infantry division, had previously participated in the Soviet invasion of Afghanistan.[15] In 1992, there were three units associated with the 201st MRD: in Dushanbe (the 92nd Motor Rifle Regiment), in Qūrġonteppa (the 191st Motor Rifle Regiment), and in Kūlob (the 149th Guards Motor Rifle Regiment).[16]

Besides the 201st MRD, Russian military presence in Tajikistan consisted of a regiment of Air Defense Forces and Border Guard Forces deployed at the borders of Tajikistan with China and Afghanistan.[17] The border guards had previously been subordinated to the Soviet Committee for State Security (more commonly known as the KGB), but they had been reassigned to a special command in the Ministry of Defense of the Russian Federation by the time the civil war started.[18] After the dissolution of the Soviet Union, Russia did not have a clear understanding of the role that its troops in Tajikistan were to play. With large numbers of Red Army forces withdrawing from Eastern Europe to Russia, moving the 201st and the other forces from Tajikistan to Russia was not an attractive option.[19]

By May 1992, Tajik internal security forces had largely disintegrated. In the absence of a national army to control the mounting unrest in the

---

[14] Political Economy Research Institute, undated.

[15] Pirseyedi, 2000, p. 48.

[16] Tim Epkenhans, *The Origins of the Civil War in Tajikistan: Nationalism, Islamism and Violent Conflict in Post-Soviet Space*, Lanham, Md.: Lexington Books, 2016, p. 167.

[17] Pirseyedi, 2000, p. 48; Epkenhans, 2016, p. 244.

[18] Rubin, 1993, p. 73.

[19] Pirseyedi, 2000, pp. 48–49.

country,[20] the sitting government of Tajikistan asked for Russian support as clashes with the opposition intensified in the spring of 1992.[21] Because Moscow was preoccupied with addressing several impending domestic political matters ensuing from the dissolution of the Soviet Union,[22] the Russian military units in Tajikistan did not receive specific operational instructions except for remaining neutral in the conflict.

Despite this, Russian troops gradually became involved in the civil war on the side of the government.[23] Although neither Moscow nor the Russian troops present in Tajikistan instigated the outbreak of internal violence,[24] their support of the government and former Communist Party elites allowed the latter to maintain their grip over Tajik politics.[25]

In our analysis of the drivers of the gradual escalation of Russian involvement in the civil war in Tajikistan, we focus on the period between May 5, 1992, the official date when the war started, and the end of July 1993, when, immediately after an attack that killed 24 Russian border guards,[26] 10,000 additional Russian troops were sent as reinforcements to guard the 1,400-km Tajik-Afghan border.[27] Although maintaining an official policy of military neutrality during this period, Russia's role changed gradually

---

[20] Michael Orr, "The Russian Army and the War in Tajikistan," in Mohammad-Reza Djalili, Frédéric Grare, and Shirin Akiner, eds., *Tajikistan: The Trials of Independence*, New York: Routledge, 1998.

[21] Catherine Pujol, "Some Reflections on Russian Involvement in the Tajik Conflict, 1992-1993," in Mohammad-Reza Djalili, Frédéric Grare, and Shirin Akiner, eds., *Tajikistan: The Trials of Independence*, New York: Routledge, 1998.

[22] Epkenhans, 2016, p. 288.

[23] Epkenhans, 2016, p. 167.

[24] Rajan Menon, "After Empire: Russia and the Southern 'Near Abroad,'" in Michael Mandelbaum, ed., *The New Russian Foreign Policy*, New York: Council on Foreign Relations, 1998, pp. 144–148.

[25] Pirseyedi, 2000, pp. 48–49.

[26] This incident is treated as a distinct conflict in our data set because of the involvement of Afghanistan's government, making it an interstate dispute rather than an intervention in a civil conflict.

[27] Clark, 1993, p. 15.

from bystander to combatant.[28] Russia suffered its first military casualty on May 6, 1992, after the official date of the onset of the civil war, but the gradual, drawn-out nature of the Russian intervention across several months render it difficult to pinpoint an exact date or specific event when this case qualified as a militarized conflict.[29]

During this period, the Russian government vacillated in its response to the outbreak of the Tajik Civil War. As a result of divisions within the Kremlin at the time, Russia did not have a coherent policy toward Central Asia (and its near abroad) in general nor vis-à-vis the conflict in Tajikistan, specifically.[30] Both the government and the population feared that Russia would get inadvertently "dragged into another Afghan-like quagmire"; Russian officers deployed in Tajikistan at the time "expressed doubts about their mission" and were unclear about whose interests they were expected to protect.[31]

Russia was also reluctant to assume sole responsibility for stabilizing the situation in Tajikistan, and in July 1992 asked the Kazakh, Uzbek, and Kyrgyz governments to provide troop reinforcements to defend the southern border of the newly formed Commonwealth of Independent States (CIS) with Afghanistan.[32] It was not until April 1993 that the Russian government finally articulated a fully fledged strategy addressing the near abroad when Foreign Minister Andrei Kozyrev, according to Dov Lynch, "affirmed that Tajikistan, along with the whole of Central Asia, represented a Russian 'zone of special responsibility and interest.'"[33] In this vein, between April and July 1993, Russia backed up this approach by transferring weapons to the Tajik government and committing additional troops to secure the Tajik-

---

[28] Clark, 1993, p. 15; Orr, 1998.

[29] The first Russian military casualty occurred when a car transporting a 201st MRD officer came under attack from the opposition forces, killing both driver and officer. For details, see Epkenhans, 2016, p. 289.

[30] Dov Lynch, *Russian Peacekeeping Strategies in the CIS: The Cases of Moldova, Georgia, and Tajikistan*, London: Royal Institute of International Affairs, 2000, p. 48.

[31] Clark, 1993, p. 15.

[32] Pujol, 1998.

[33] Lynch, 2000, p. 158.

Afghan border.[34] Furthermore, according to Lynch, Moscow became "more involved in Central Asian security arrangements" as a result of the 1993 attack on Russian border guards on the Afghan-Tajik border, which had an impact on Russian strategy toward the region.[35]

Drivers

Summary of Findings

Gradually gaining momentum over the span of 15 months between May 1992 and July 1993, Russia's military intervention in the Tajik Civil War had seven main drivers:

1. Tajikistan's location in geographic proximity to the Russian Federation and its status as a former Soviet republic
2. the presence of Russian military units and border guards on the territory of Tajikistan when the civil war started
3. the presence of a significant ethnic Russian community in Tajikistan
4. the external threat that Islamic extremism posed to Russia and other Central Asian governments, and the fear of domestic instability becoming contagious and spreading from Tajikistan to internally destabilize Russia
5. acute Russian uncertainty about the future in the context of the recent dissolution of the Soviet Union
6. reputational costs for Russia's nonintervention
7. domestic political instability within Russia.

Some of these drivers—geographic proximity, presence of Russian military units and ethnic Russians—can be considered structural factors; others (the rising threat of Islamic fundamentalism, acute Russian uncertainty about the future, reputational costs, and domestic political instability inside Russia) were proximate drivers with an impact on Russia's intervention in the civil

---

[34] Lynch, 2000, pp. 158–159.

[35] Lynch, 2000, p. 160.

war. However, their impact was not always unidirectional; although the rising threat of Islamic fundamentalism, acute Russian uncertainty about the future, and concern about incurring reputational costs precipitated the intervention, domestic political instability had a hindering effect on the manner in which Russia initially responded to the outbreak of the Tajik Civil War and translated into a gradual escalation of Russian involvement.

As the sections in this appendix will present in detail, the gradual escalation in the number of Russian troops deployed to Tajikistan reflected the continued rise in Russian commitment, driven by the first six factors identified above in the context of increased consensus among Russian policymakers and by Moscow's gradual development of a consistent strategy toward the region during the second half of 1992 and continuing in 1993.

Geographic and Territorial Drivers
Geographic Drivers: Former Soviet Republic Status
The evolution in Russian response and gradual increase in the number of Russian troops deployed to secure the Tajik-Afghan border during this period was associated with changes in Russian strategic thinking about the near abroad. Repeated public acknowledgements in 1993 on the part of Russian officials reaffirmed the strategic importance of the country, stemming from its geographic proximity to Russia and its belonging to the near abroad, where Russia wanted to continue projecting power and maintaining influence. At the time, President Yeltsin publicly spoke of Tajikistan's southern border as, effectively, being Russia's southern border.[36] The July 13, 1993, attack on Russian border guards forced Moscow to adopt a tougher stance toward the conflict in Tajikistan. Less than a week later, a senior Russian official proclaimed that "[d]efending the Tajik section of the border is defending the backbone of Russian security. The Russian border guards hold the key to Russian and CIS security."[37] Such statements reflect the geopolitical importance that Tajikistan and its borders held in Russian military thinking at the time and the Russian commitment to defend the Tajik-Afghan border.[38]

---

[36] Pirseyedi, 2000, p. 51.

[37] Pujol, 1998.

[38] Lynch, 2000, p. 29.

Nature of the Relationship Between Russia and the Opposing State

Presence of Russian Military Installation

At the time when the civil war broke out in May 1992, the 201st MRD had a nominal strength of 12,000,[39] and its presence in Tajikistan represented one of the main drivers behind Russia's intervention in the conflict. As violence intensified in spring 1992, in the absence of a national army, the Tajik government asked for Russian military support. On May 5, 1992, Tajik President Nabiyev declared a state of emergency and had a phone call with Russian President Yeltsin. In the aftermath of the call, the 201st MRD was deployed to guard sensitive facilities in Dushanbe while being directed to not intervene in the confrontations on the ground.[40]

In May 1992, the Tajik government asked for an increase in the scope of 201st MRD's involvement in the conflict, seeking support to protect critical infrastructure—including the Norak Hydroelectric Dam and the Yovon Electro-Chemical Plant. In July 1992, Russian Defense Minister Pavel Grachev and Russian Chief of Defense Yevgeniy Shaposhnikov (who, at the time, was also commander in chief of the short-lived CIS Joint Armed Forces) agreed to the Tajik government's request. Both conscripted and officer rank personnel with the 201st MRD, as a result of their involvement in the conflict, were caught in crossfire between the government and opposition forces in the following months.[41]

---

[39] According to several accounts, the division had been operating under strength since at least 1990, with Slavic officers needing to be redeployed to Russia and local Tajiks hesitating to respond to conscription calls. See Epkenhans, 2016, pp. 167, 289. According to Pujol, the division had operated below strength by 40 percent for several years and had around 6,000 men (Pujol, 1998).

[40] Epkenhans, 2016, p. 277.

[41] Epkenhans, 2016, pp. 289–290. There is debate about the precise level of involvement of the 201st MRD in the violence that occurred in the summer and fall of 1992. Official statements tended to downplay the division's involvement in the conflict, although there were claims that the 201st MRD was actively involved in fighting the opposition in 1992 and 1993 (Epkenhans, 2016, p. 291; Pirseyedi, 2000, p. 49; and Flemming Splidsboel-Hansen, "The Outbreak and Settlement of Civil War: Neo-realism and the Case of Tajikistan," *Civil Wars*, Vol. 2, No. 4, 1999).

After May 1992, Tajik opposition militia members increased the number of crossings into Afghanistan to procure weapons and ammunition. Daily skirmishes took place as Russian border guards attempted to contain these illegal border crossings.[42] In September 1992, the Russian Ministry of Defense decided to send troops as reinforcements to make up for the desertions that had taken place and for losses resulting from heavy fighting.[43] Some 1,000 border guards (mainly Russian, but also including troops from the other Central Asian states) were deployed in September 1992 along the Amu Darya river, bringing the total to 3,500.[44] Additionally, Spetsnaz commandos were sent into Tajikistan in fall 1992, prior to the arrival of substantial traditional military reinforcements in 1993 in the context of the CIS peacekeeping force.[45]

A November 1992 summit of Central Asian and Russian defense ministers led to the creation of a Russia-led CIS peacekeeping force to stabilize Tajikistan, which also marked the deepening of Russian involvement, including both military and economic aid to the government.[46] The 201st MRD was given a peacekeeping mandate.[47] But, as Barnet Rubin aptly remarked, "[t]here was, however, no peace to keep. The force turned into a prop for the new regime that subsequently took power in Dushanbe."[48] After the July 13, 1993, cross-border attack by opposition forces on Russian border guards, some 10,000 more troops were deployed as reinforcements

---

[42] Brown, 1998.

[43] By August 1992, "the 191st MRR [Motor Rifle Regiment] in Qūrġonteppa, which had a nominal strength of 2,000 men, was down to 52 servicemen, 49 of them officers" (Epkenhans, 2016, p. 292). For more details, see Epkenhans, 2016, pp. 291–292.

[44] Rubin, 1993, p. 80.

[45] Epkenhans, 2016, p. 291.

[46] Lynch, 2001, pp. 55–56; Rubin, 1993, p. 81.

[47] Lynch, 2000, p. 77.

[48] Rubin, 1993, p. 80.

on the Tajik-Afghan border.[49] The attack forced Moscow not only to adopt a stronger stance toward the conflict in Tajikistan but also to increase its involvement in security arrangements in the near abroad. The CIS peace-keeping force was finally deployed in autumn 1993.

Presence of Russian Compatriots

At the beginning of the civil war in May 1992, approximately 388,000 ethnic Russians lived in Tajikistan, representing slightly less than 10 percent of the country's total population. Together with the Uzbek minority, the Russians were perceived to represent "the better off" groups in the impoverished country.[50] Kevlihan and Menon both argue that, in its early stages, the Russian military intervention in the Tajik Civil War was strongly focused on securing the capital and providing protection to ethnic minorities, including the large Russian community in Dushanbe.[51]

Pirseyedi makes a similar argument regarding the "plight of the Russian minority" as a driver of the Russian intervention in the internal conflict. The increased involvement of Russian military forces into the civil war resulted in claims that Russian troops attacked opposition bases, confiscated weapons, and killed innocent civilians, rendering the Russian military forces to be perceived by many Tajiks as an occupation force.[52] In response, attacks were directed not only against Russian military personnel and installations, but also against ethnic Russians, and anti-Russian rhetoric increased during this time. By April 1993, fearing rising nationalism and Islamic extremism (and as a result of attacks against them), some 300,000 ethnic Russians had fled Tajikistan.[53] The escalation in violence against ethnic Russians resulted in a larger Russian military commitment to the defense of the capital.

---

[49] Menon, 1998, pp. 144–148; Rubin, 1993, p. 86.

[50] Anna Matveeva, *The Perils of Emerging Statehood: Civil War and State Reconstruction in Tajikistan: An Analytical Narrative on State-Making*, London: Crisis States Research Centre, Working Paper No. 46, March 2009, p. 28.

[51] Rob Kevlihan, "Insurgency in Central Asia: A Case Study of Tajikistan," *Small Wars & Insurgencies*, Vol. 27, No. 3, 2016, pp. 421–422; Menon, 1998, pp. 144–148.

[52] Pirseyedi, 2000, p. 49.

[53] Matveeva, 2009, p. 28; Rubin, 1993, p. 80.

Russian Threat Perceptions and Status Concerns

Russia Perceives Increased External Security Threats

Russia feared the expansion of Islamic fundamentalism from Afghanistan—where the Islamic militants were making advances—into Central Asia and subsequently into Russia. Under these circumstances, Russian military presence and involvement in Tajikistan to stabilize the country and secure the border with Afghanistan were necessary steps to prevent the spillover of Islamist fighters into Central Asia and into Russian territory.[54] Many governments in the region, especially Moscow, perceived Tajikistan as representing "a bridgehead" for Islamist forces "into Central Asia and from there into Russia."[55]

The fear of Islamic fighters spreading their violent activities beyond Afghanistan into Tajikistan and farther into Central Asia was rooted in the role that Afghanistan played during the Tajik Civil War. Afghanistan was an important source of weapons and fighters, and it provided sanctuary for Tajik opposition forces. Not only did some 12,000 opposition fighters find refuge across the border in Afghanistan, but Tajik refugee camps in Afghanistan transformed into highly militarized recruitment and training centers for opposition fighters. Together with Afghan fighters and other volunteers from Middle Eastern countries, they carried out raids across the border into Tajikistan.[56] Russia considered its military presence in the country "to be vital for countering the threat of the spillover of Tajik and Afghan conflicts into Central Asia" and necessary to contain the spread of Islamic fundamentalism.[57]

---

[54] Menon, 1998, pp. 144–148; Pujol, 1998; and Pirseyedi, 2000, p. 51.

[55] Mohammad-Reza Djalili and Frédéric Grare, "Regional Ambitions and Interests in Tajikistan: The Role of Afghanistan, Pakistan and Iran," in Mohammad-Reza Djalili, Frédéric Grare, and Shirin Akiner, eds., *Tajikistan: The Trials of Independence*, New York: Routledge, 1998.

[56] Pirseyedi, 2000, p. 54.

[57] Pirsevedi, 2000, p. 51.

Russia Perceives Increased Acute Uncertainty About the Future and Reputational Costs for Nonintervention

After the dissolution of the Soviet Union, Russia's uncertainty about the future, the role that the country would play in international politics, and the turn that regional dynamics would take all influenced Moscow's attempts to maintain influence in its neighborhood. The prospect of a domino effect with the other Central Asian governments suffering the fate of Tajikistan if the latter were to fall prey to Islamic extremism seemed real to many. Few in Moscow had any certainty about the future of regional security, the nature of the Tajik government, the role of Islam in post-Soviet countries, or any number of other key considerations for planning.

Moscow lacked a coherent policy vis-à-vis the immediate neighborhood in the early stages of the Tajik Civil War, but Russia's ambitions to continue projecting power and maintaining influence in Central Asia had not dissipated. By the end of 1992, it had become clear in Moscow that Russia had to intervene in the Tajik Civil War by supporting the Tajik government and consolidating the power of former Communist Party elites with strong links to Russia as a way to protect its interests in the region.[58] Following this line of thinking, had Russia not intervened militarily in the conflict in Tajikistan, a signal of weakness and decline would have been sent to Russia's allies in the region and to its adversaries. Russian abandonment of Tajikistan was likely to be perceived by both groups as "abandoning a claim to a sphere of influence in [Central] Asia."[59] The need to avoid such a perception justified Russia's continued military presence in Tajikistan, even as the costs associated with Russia's military intervention and economic commitment to Tajikistan kept rising.[60] Russia thus aimed to preserve a reputation for resolve to deter any potential adversaries—specifically, Islamic extremists—from taking their fight closer to Russia's borders.[61]

---

[58] Pujol, 1998.

[59] Gregory Gleason, "Why Russia Is in Tajikistan," *Comparative Strategy*, 2001, Vol. 20, No. 1, pp. 85–87.

[60] Gleason, 2001, pp. 81–82.

[61] Pirseyedi, 2000, p. 51.

Domestic Drivers

Russia Experiences Domestic Political Instability

With its attention directed toward internal matters and establishing relations with the broader international community rather than the former Soviet states, the Kremlin's response to the civil war in Tajikistan was mostly cacophonous and incoherent during the spring and part of the summer of 1992.[62] Lynch referred to Russia's strategy toward the near abroad during 1992 as an "empty vessel, characterized by ill-defined generalities."[63] The shift that occurred in Russian foreign policy in the second half of 1992 represented nothing more than "a gradual filling of the vessel" during which the Russian Ministry of Foreign Affairs' approach to the former Soviet countries evolved from "benign neglect to deep engagement."[64]

In this vein, after August 1992, Moscow's position toward the civil war in Tajikistan started to become less contradictory; by November 1992, the Kremlin began articulating a coherent strategy toward the region and prioritizing the countries of the former Soviet Union in its security policy.[65] As Russian strategic thinking about the region became more consistent, troop deployments to Tajikistan increased.

Conclusion

Russia's military intervention in the Tajik Civil War took place gradually between May 1992 and July 1993. The vacillation in that period reflected the lack of a coherent strategy toward the region in the early months of the conflict, mainly because of the domestic political instability that Russia experienced after the Soviet Union collapsed. Geographic proximity was clearly a major structural driver of the outcome, although the presence of

---

[62] Lynch, 2000, p. 156. For example, in June 1992, President Yeltsin refused to open a new front in Tajikistan while, at the same time, Russian Marshal Shaposhnikov was publicly discussing the option to send troops to intervene in the internal war (Pujol, 1998).

[63] Lynch, 2000, p. 37.

[64] Lynch, 2000, p. 37.

[65] Lynch, 2000, p. 6.

Russian military units and border guards and of a significant ethnic Russian community in Tajikistan were liabilities as much as assets; when they came under fire, Moscow felt compelled to respond. Other important drivers of Russia's involvement were the fear of rising Islamic extremism and of internal violence spreading to Central Asia and spilling over into Russia, acute Russian uncertainty about the future, and status concerns associated with the potential costs of nonintervention.

Militarized Conflict Case Study: Syrian Civil War, 2015–ongoing

Russia's intervention in September 2015 in Syria's civil war surprised many observers. Most had assumed that Moscow had neither the interest nor the capabilities to conduct a combat operation beyond post-Soviet Eurasia. Although it is true that Russia had been assisting the regime of Bashar al-Assad since the outbreak of hostilities in Syria in March 2011, the air operation marked a distinct escalation of its involvement. This appendix seeks to assess why Moscow's involvement escalated to the point of a militarized conflict.

Russia was involved in the Syrian Civil War in various forms before 2015. It supplied some materiel to Damascus as Assad suffered losses on the battlefield in 2012.[1] Until 2014, however, most of the deliveries were arranged under commercial contracts, not through direct military assistance.[2] Russia provided modest economic support as well.[3]

By 2015, the Assad regime's forces had lost control over major swaths of the country, including the economic center of Aleppo. In many areas, regime forces were displaced by a variety of rebel groups, some of which were backed by U.S. allies—Saudi Arabia, Turkey, Qatar, and the UAE. Several of these groups became interlinked with such extremist elements as Jabhat al-

---

[1] Ian Black and Chris McGreal, "Syria: U.S. Accuses Russia of Sending Attack Helicopters," *The Guardian*, June 12, 2012; and Jonathan Saul, "Exclusive: Russia Steps Up Military Lifeline to Syria's Assad—Sources," Reuters, January 17, 2014.

[2] Miriam Elder, "Syria Will Receive Attack Helicopters From Russia, Kremlin Confirms," *The Guardian*, June 28, 2012.

[3] Dafna Linzer, Jeff Larson, and Michael Grabell, "Flight Records Say Russia Sent Syria Tons of Cash," *ProPublica*, November 26, 2012.

Nusra, the al-Qaeda franchise in Syria. Meanwhile, ISIS forces were also making significant inroads. ISIS's spectacular offensive success in summer 2014 in both Syria and Iraq worried many in Moscow. The group's declared objective of establishing a caliphate resonated with some of Russia's disenchanted Muslim population and those in neighboring states.[4] Thousands of Russian-speaking fighters, from Russia itself and from post-Soviet Central Asia and the Caucasus, joined the ranks of ISIS and other jihadi groups.[5]

In May 2015, ISIS took control of Palmyra, the strategic and storied city in the south of the country. At the same time, groups led by Jabhat al-Nusra dealt several major defeats to regime forces in northwest Syria.[6] At some point in spring or early summer 2015, Moscow saw these developments portending the decisive defeat of the Assad regime.[7] The Kremlin apparently concluded that it needed to use military force to prevent that outcome.

In addition to the Syrian army, Moscow worked with the key ground forces backing the regime—specifically, Iranian forces and Iranian-backed Hezbollah. Iranian Revolutionary Guards Corps General Qassem Soleimani reportedly visited Moscow on July 24–26, 2015, to arrange coordination between Russian air power and Iranian-led ground forces. The formal agreement for the deployment of a Russian air contingent was signed with the Syrian government on August 25, 2015. Over the course of September, many of the fixed-wing aircraft transited through Iran and Iraq to the new Russian base at Khmeimim, near Latakia. By the end of September, the Russian force would consist of 32 fixed-wing aircraft and 17 helicopters.[8]

---

[4] Fiona Hill, "Putin Battles for the Russian Homefront in Syria," Brookings Institution, May 23, 2016.

[5] Soufan Group, *Foreign Fighters: An Updated Assessment of the Flow of Foreign Fighters into Syria and Iraq,* New York, December 2015.

[6] A. V. Lavrov, "Khod boevykh deistvii v 2011–2015 godakh," in M. Yu. Shepovalenko, ed., *Siriiskii rubezh,* Moscow: Tsentr analiza strategii i tekhnologii, 2016; and John W. Parker, *Putin's Syrian Gambit: Sharper Elbows, Bigger Footprint, Stickier Wicket,* Washington, D.C.: Center for Strategic Research, Institute for National Strategic Studies, National Defense University, Strategic Perspectives No. 25, July 2017, pp. 10–12.

[7] For a Russian analysis of the situation, see, for example, Maksim Yusin and Sergei Strokan', "Ni mira, ni Pal'mira," *Kommersant,* May 22, 2015.

[8] Lavrov, 2016.

Putin addressed the UN General Assembly on September 28 and called for a cooperative global effort to fight the scourge of terrorism in Syria. Two days later, Assad formally asked Russia for military assistance and the Federation Council, Russia's upper house of parliament, formally granted permission for the Russian bombing campaign, which began that day. The Russian intervention has largely proven a success. As of mid-2020, Assad has reestablished government control over the vast majority of population centers. None of the remaining rebel groups poses a threat to his rule. Russia's intervention dramatically turned the tide of the civil war and also cemented Moscow's role as a regional power broker.

Drivers

Summary of Findings

There were four central drivers of the escalation of Russia's involvement in the Syrian Civil War:

1. Russia's increased acute uncertainty about the future generated by the situation on the ground in Syria
2. Russian perceptions of increased external security threats
3. Russian dissatisfaction with its role in the international system
4. Russian perceptions of reputational costs for nonintervention.

First, the situation on the ground in Syria fostered acute increased uncertainty about the future among Russian decisionmakers. The possibility that Assad would fall within months, perhaps even weeks, barring an external military intervention and that he would likely be replaced by a regime hostile to Russian interests—or by no regime at all, a situation that Moscow saw as potentially empowering extremist elements—created a sense of urgency about the need to intervene.

Assad's potential ouster was associated with two additional drivers: external threats and reputational costs for nonintervention. Moscow believed that Assad's fall would legitimize what Russia sees as the U.S. policy of ousting governments that do not comply with Washington's wishes, a policy that Moscow believes could be applied to Russia itself in the future. Russia's leaders might have seen intervention in Syria as a sort of forward defense of the homeland. Additionally, the Kremlin saw links between extremists in Russia and their counterparts in Syria that created a terrorist threat. Moreover, had Moscow not intervened, it risked losing its only client in the Middle East and its sole remaining military facilities there—and, thus, its regional clout.

Finally, the escalation of Russia's involvement in Syria was also driven by Russia's dissatisfaction with its place in the international system. Following the annexation of Crimea and the invasion of eastern Ukraine in 2014, the West had attempted to isolate Russia diplomatically. Intervening in the Syrian Civil War would, Moscow believed, compel the West to deal with Russia and would represent a breakdown of the West's attempted diplomatic blockade.

In the following sections, we present a detailed analysis of these four drivers of escalation, but we also discuss why the presence of a Russian military installation in Syria was not a driver in this particular case, especially when we consider the role that this factor played as a driver of escalation in the Tajik case.

Nature of the Relationship Between Russia and the Opposing State

Presence of Russian Military Facility

We begin with a brief treatment of a factor that was present in this case but did not drive escalation: the presence of Russian military installations in Syria. Although Russia's relationship with Syria might have been relatively less important than relationships with partners in other regions, Syria did offer Moscow a limited power projection platform in the Middle East. At the time of the intervention, Russia maintained a small naval facility at Tartus that hosted a small contingent of sailors and support personnel and could be used as a refueling point. The facility, however, had no capacity to host large warships.[9] In addition, the Russian and Syrian militaries also reportedly operated two joint signals intelligence facilities, which were seized by rebels in 2014.[10] Ultimately, the physical assets themselves did not create escalatory

---

[9] Frank Gardner, "How Vital Is Syria's Tartus Port to Russia?" *BBC News*, June 27, 2012.

[10] Inna Lazareva, "Russian Spy Base in Syria Used to Monitor Rebels and Israel Seized," *The Telegraph*, October 8, 2014.

pressure for Russia. Losing the signals intelligence (SIGINT) sites did not in itself seem to drive action in Moscow, although it underscored the severity of the rebellion's challenge to the regime. And maintaining Tartus was not important enough to Moscow to justify a significant military intervention. In short, although these facilities certainly facilitated the intervention, they did not appear to drive it.

Russian Threat Perceptions and Status Concerns

Russia Perceives Increased Acute Uncertainty About the Future

By mid-2015, the Syria conflict contributed to a change in Russian decisionmakers' thinking on intervention. Prior to 2015, Moscow supported the Assad regime in various ways, but the regime had experienced setbacks by spring and summer of that year that suggested its collapse might be imminent. Assad's forces had been losing on the battlefield since 2013 and ceding ground to ISIS, the Kurds, the Free Syrian Army, Nusra-led groups, and others. By spring of 2015, two events likely contributed to Moscow's view that regime collapse was imminent: ISIS's seizure of Palmyra in May and the presence of a rebel coalition in northwest Syria.[11] As Russian President Putin emphasized, Syrian "statehood" could collapse—and, without decisive action, the situation could lead to "the complete Somalization of [Syria], the complete degradation of statehood."[12] He also warned that Russia could not allow a situation similar to the ones that occurred in Libya and Somalia to emerge in Syria.[13] Reflecting on the decision to intervene, Russian leaders have argued that if the decision to do so had not been made, the Assad regime would have fallen in months, perhaps even weeks. For example, as Valery Gerasimov, chief of the Russian General Staff, stated, "in 2015, just over 10 percent of [Syrian] territory remained under government control.

---

[11] Parker, 2017, pp. 10–12.

[12] Vladimir Smirnov, "'Zachem nam mir bez Rossii?': Vladimir Putin o vozmozhnosti primeneniya yadernogo oruzhiya, terrorizme i situatsii v Sirii," *RT na ruskom*, March 7, 2018; and President of Russia, "Meeting of the Valdai International Discussion Club," Sochi, October 18, 2018.

[13] "Putin schitaet nedopustimym povtorenie v Sirrii 'liviiskogo' stsenariya," *RIA Novosti*, June 2, 2017.

A month or two more, by the end of 2015, and Syria would have been completely under ISIS [rule]."[14] In the Kremlin's view, Assad's fall could lead to two outcomes that both had negative implications for Russian security: Transnational terrorist groups would achieve victory, and Western-led regime change would be legitimized. We discuss both of these next.[15]

Russia Perceives Increased External Security Threats

Russian leaders viewed the potential for a transnational terrorist victory and a successful Western-led regime change effort as significant external security threats. Viewing the situation through the lens of Russia's experience with Chechen separatists and the Soviet Union's experience in Afghanistan, the Kremlin believed that Assad's defeat would have led to significant instability.[16] In addition, Russian decisionmakers saw similarities between the situation in Syria and the aftermath of Western efforts to oust Muammar Qaddafi from power in Libya, where ISIS flourished. The Kremlin believed that Assad's defeat would amount to a Sunni extremist victory. "ISIS would have continued to gather momentum and would have spread to adjacent countries," Gerasimov claimed. "We would have had to confront that force on our own territory," as ISIS forces would operate "in the Caucasus, Central Asia, and the Volga region [of Russia]."[17]

Although Gerasimov might have overstated his case, the Kremlin did appear to link the success of extremist groups in Syria to an increased risk of domestic terrorism, particularly in regions with significant Muslim populations. Underscoring this point, Putin declared the need to "take the initiative and fight and destroy the terrorists in the territory they have already captured rather than waiting for them to arrive on our soil."[18]

---

[14] Viktor Baranets, "Nachal'nik genshtaba vooruzhennykh sil Rosssii general armii Valerii Gerasimov: 'My perelomili khrebet udarnym silam terrorizma,'" *Komsomol'skaya pravda*, December 26, 2017.

[15] Samuel Charap, Elina Treyger, and Edward Geist, *Understanding Russia's Intervention in Syria*, Santa Monica, Calif.: RAND Corporation, RR-3180-AF, 2019, p. 4.

[16] Charap, Treyger, and Geist, 2019, p. 4.

[17] Baranets, 2017.

[18] President of Russia, "Meeting with Government Members," transcript, Novo-Ogaryovo, Moscow Region, September 30, 2015b.

Traditionally, Moscow has viewed what it believes to be Western, and particularly U.S., regime change efforts around the world as direct threats to its national security. In the Syrian case, Russian decisionmakers also viewed the potential overthrow of Assad by a U.S.-backed rebel coalition as legitimizing the West's practice of regime change. In the Kremlin's perception, NATO's ouster of Qaddafi in Libya, the U.S. invasion of Iraq to overthrow the Saddam Hussein regime, the coalition invasion of Afghanistan, and the NATO air campaign in Serbia all represent relevant examples of a long-held Western practice of regime change and interference in the domestic affairs of other countries.

Beyond those examples, the Kremlin also contended that the United States pursued the same objective nonkinetically by supporting or instigating several color revolutions in former Soviet states and the Arab Spring. Putin has claimed that Western regime-change efforts are the source of significant instability in the international system: "Unilateral dictates and forcing one's own political framework [onto other states] produces exactly the opposite [of the intended result]: instead of conflict settlement, escalation; instead of sovereign, stable states, a growing expanse of chaos."[19] Animated by this belief, Moscow sought to prevent regime change in Syria as a means to secure regime stability at home. A goal of the Russian intervention was to discredit a tactic that might be used at a later point against the Russian government.[20]

Russia Experiences Dissatisfaction with Its Place in the International System

As noted previously, we consider Russia to have been dissatisfied with its place in the international system since 2007. However, Russia's dissatisfaction was even more acute in 2015 after the United States and its allies sanctioned Russia both economically and politically, removing it from international bodies and shunning Moscow diplomatically. This exacerbated all the preexisting resentment and antihegemonic instincts in Moscow. Russian leaders believed that intervening in Syria could not merely prevent a loss of influence in the region but rather increase Russia's leverage and allow it to

---

[19] President of Russia, "Meeting of the Valdai International Discussion Club," excerpts of transcript, Sochi, October 24, 2014c.

[20] Charap, Treyger, and Geist, 2019, pp. 5–6.

once again play a prominent role in Middle Eastern, perhaps even international, politics.[21] The U.S. and EU sanctions and efforts to, as U.S. President Barack Obama put it, "isolate" Russia directly challenged Moscow's conception of its own great power status.[22] In March 2014, the G-8 removed Russia as a participant and NATO suspended the NATO-Russia Council, a body for Russia and NATO member states to discuss matters of mutual concern. Meanwhile, the EU and United States canceled all presidential-level engagement with Russia while U.S. agencies cut off all nonessential cooperation with their Russian counterparts. President Obama described Russia publicly as a "regional power . . . acting out of weakness."[23]

Under this sustained geopolitical press, Syria was intended to be used "as a bargaining chip in relations with the West," according to one Russian analyst.[24] At the beginning of the intervention, Putin evoked the U.S.-UK-Soviet alliance when proposing a grand counterterrorism coalition at the UN General Assembly.[25] In addition, Russian military officials have regularly sought official coordination with the United States on counterterrorism operations in Syria.[26]

Reputational Costs for Russia for Nonintervention

Moscow feared that, had it not intervened in Syria, it might have faced reputational costs: Without taking action, Russia risked losing its status as an external power with influence in the Middle East. Moreover, given Moscow's opposition to what it views as a Western policy of regime change, deciding not to intervene in Syria might have led other states in the region and elsewhere to question Russia's credibility. At the time of the intervention, Moscow's only significant partner in the region was the government in

---

[21] Charap, Treyger, and Geist, 2019, p. 7.

[22] Zeke J. Miller, "Obama: U.S. Working to 'Isolate Russia,'" *TIME*, March 3, 2014.

[23] Scott Wilson, "Obama Dismisses Russia as 'Regional Power' Acting out of Weakness," *Washington Post*, March 25, 2014.

[24] Cited in Charap, Treyger, and Geist, 2019, p. 7.

[25] Vladimir Putin, "70th Session of the UN General Assembly," transcript, President of Russia, September 28, 2015a.

[26] Charap, Treyger, and Geist, 2019, p. 8.

Syria; no other Middle Eastern country hosted Russian military and intelligence facilities, and Russia and Syria had a strong tradition of military-to-military links.[27] Not standing up for its ally would have made Russia akin to the United States in its refusal to back Egyptian President Hosni Mubarak during the popular rebellion against his rule. Russia's ambassador to the UN at that time, Vitaly Churkin, said in 2015, "Being a Russian diplomat, for us, if you have good relations with a country, a government, for years, for decades, then it's not so easy to ditch those politicians and those governments because of political expediency."[28]

As the conflict intensified, Moscow began to see geopolitics as the key driver of Western and regional Sunni-led states' policies; this, in turn, spurred further interest in defending the Assad regime and contributed to pressures on the Russian government to escalate its involvement. Russia set its own red line of sorts as the situation evolved. Although Moscow might have been indifferent to who was in power in Damascus before the war (as long as they were friendly to Russian interests), Russia became more invested in assisting Assad himself to maintain power as Russian leaders began to view the conflict through a geopolitical lens. In 2012, Russian Foreign Minister Lavrov said the war was "reformatting the geopolitical map of the Middle East, where different players are trying to secure their own geopolitical positions." He claimed that the opposition's international backers "openly [said] that it is necessary to deprive Iran of its closest ally."[29]

By framing the conflict as a geopolitical issue, Moscow raised the potential reputational costs associated with not intervening to prevent Assad's fall. Through this lens, Russian leadership concluded that allowing the collapse of the Assad regime could lead future regional partners not only to

---

[27] Since then, Egypt might have allowed Russia to access its air base outside Sidi Barrani, though nothing confirmed publicly; Russia might be operating unmanned systems there in support of its favored factions in Libya (Phil Stewart, Idrees Ali, and Lin Noueihed, "Exclusive: Russia Appears to Deploy Forces in Egypt, Eyes on Libya Role—Sources," Reuters, March 13, 2017).

[28] Cited in Colum Lynch, "Why Putin Is So Committed to Keeping Assad in Power," *Foreign Policy*, October 7, 2015.

[29] Sergei Lavrov, "Za i PROtiv: Sergei Lavrov o vneshnepoliticheskikh vragakh, o vozmozhnoi voine mezhdu SShA i Iranom i mnogom drugom," *Rossiiskaya Gazeta*, October 24, 2012.

doubt Russia's willingness to stand by allies during a crisis but also to question Russian resolve to pursue its geopolitical interests.

Conclusion

Increased uncertainty about the future, external security threats, and status and reputational-driven issues drove escalation in this case. Acute uncertainty about the future among Russian decisionmakers—fearing that Assad would fall imminently and be replaced by a regime hostile to Russian interests or by complete state collapse (which, in turn, would allow extremist elements to take power)—created a sense of urgency about the need for decisive action. Overlaying this proximate cause for intervention were (1) Moscow's belief that Assad's fall would legitimize what it believes to be a Western policy of regime change, which could be used against Russia, and (2) fears of increased terrorist links between extremists in Russia and their counterparts in Syria. Russian leaders saw the situation in Syria as contributing to or exacerbating direct threats to the Russian homeland. More broadly, these national security issues were seen through the lens of Russia's concerns about its status in the international system and potential reputational costs for not intervening. When deciding to escalate, Moscow took into account the West's attempts to isolate it after the 2014 annexation of Crimea and subsequent invasion of eastern Ukraine. Moscow concluded that escalation would force the West to negotiate with Russia. Had the intervention not occurred, Moscow believed it would lose its only client in the Middle East—along with its sole remaining military facilities there and thus its regional clout.

Narratives for Ukrainian and Georgian Disputes

Disputes with Ukraine

Dispute 1: 1992 Division of Control over the Black Sea Fleet

This dispute took place in the context of the collapse of the Soviet Union, when the newly independent Russia and Ukraine disagreed over which country should have control over the Black Sea Fleet based in Crimea. Russia and Ukraine each maintained that it had the right to full control over the totality of the Black Sea Fleet. Ukraine claimed that it should control the Black Sea Fleet because it was located in Crimea, a Ukrainian territory.[1] Russia originally accepted Ukraine's claim but later backtracked and asserted its unilateral right to the fleet,[2] claiming that the Black Sea Fleet's nuclear forces meant that it should be under the Russian-controlled Armed Forces of the CIS.[3] During summer 1992, negotiations failed as both countries refused to modify their respective positions.

A series of militarized incidents occurred during the dispute, further stressing Russia-Ukraine relations.[4] The dispute ended with an agreement between Russia and Ukraine, signed in Yalta on August 3, 1992, that placed the

---

[1] Felgenhauer, 1999, pp. 4–7; Kryukov, 2006.

[2] Kryukov, 2006.

[3] Felgenhauer, 1999, pp. 4–7.

[4] Felgenhauer, 1999, pp. 4–7; and "Sailors Mutiny and Take a Ship to Ukraine," *New York Times*, July 22, 1992.

Black Sea Fleet under the joint control of Russia and Ukraine. Although this agreement ended the immediate standoff between Russia and Ukraine, the larger issue of control over the Black Sea Fleet and Crimea was not resolved.[5]

Dispute 9: 1994 Cheleken Incident

In September 1993, Russia and Ukraine signed the Massandra Accords, which allowed Russia to continue basing the Black Sea Fleet and its associated infrastructure in Crimea.[6] Moscow and Kyiv agreed to the 50–50 division of the Black Sea Fleet, but Ukraine was unable to pay for its portion of the fleet and stopped doing so in December 1993.[7]

Separate from these financial tensions, a series of hostile interactions between Russian and Ukrainian sailors occurred after the 1992 dispute. These incidents involved sailors from both countries tying up and assaulting one another, and Russian sailors chasing Ukrainian ships and forcefully preventing them from docking.[8]

On Friday, April 8, 1994, the *Cheleken*, a Russian Navy hydrographic research vessel carrying $10 million in research and navigational equipment, left the Port of Odessa. Russia claimed that the ship was sailing from Odessa to another Ukrainian port,[9] but Moscow might have been attempting to move the vessel because of Ukraine's failure to provide financial support for its maintenance.[10] Meanwhile, Ukraine claimed that the *Cheleken*

---

[5] Felgenhauer, 1999, pp. 4–7; and Government of the Russian Federation, *Soglashenie mezhdu Rossiiskoi Federatsiei i Ukrainoi o printsipakh formirovaniya VMF Rossii i VMS Ukrainy na baze Chernomorskogo flota byvshego SSSR*, Moscow, August 3, 1992.

[6] Kryukov, 2006; Government of the Russian Federation, *Protokol ob uregulirovaniya problem Chernomorskogo flota*, Moscow, September 3, 1993.

[7] Richard Boudreaux, "Kiev-Moscow Pact Gives Russia Most of Black Sea Fleet: Accord: Agreement Defuses Volatile Situation. Ukraine Was Approaching a Showdown over the Disputed Armada," *Los Angeles Times*, April 16, 1994, p. 9.

[8] Lee Hockstader, "Brush with Black Sea Naval Battle Heightens Russo-Ukrainian Tensions; Warships, Fighter Jets Dispatched in Weekend Confrontation," *Washington Post*, April 11, 1994a.

[9] Hockstader, 1994a.

[10] Hockstader, 1994a; and Lee Hockstader, "Ukraine Detains Officers After Russia Grabs Ship, as Fleet Conflict Escalates," *Washington Post*, April 12, 1994b.

was not authorized to leave Odessa and that Russia was stealing the ship and its equipment.[11] In response, Ukraine scrambled four Su-15 fighters to buzz the ship and pursued the *Cheleken* with Ukrainian ships.[12] On April 9, the *Cheleken* docked in Sevastopol, placing the ship under the control of the Russian Navy.[13] Further altercations followed between Russian and Ukrainian forces on the ground, culminating when Ukrainian forces seized a Russian military base.[14]

After the seizure of the base, the two parties came to the negotiations table and agreed to allow Russia to continue basing its part of the Black Sea Fleet in Sevastopol as part of a lease agreement. Though the Russian portion of the Black Sea Fleet would remain in Sevastopol, the accord stated that the Russian and Ukrainian portions of the Black Sea Fleet would be based separately.[15]

Disputes with Georgia

Dispute 39: 2003 Deployment of Russian Troops and Equipment to Tskhinvali

Tensions escalated in the breakaway region of South Ossetia, in Georgia, over the course of four months. In October 2002, the de facto government of South Ossetia feared that the Georgian government was planning to use an anticrime operation in the Shida Kartli region as pretext for reintegrating South Ossetia into internationally recognized Georgian territory. As a result, the South Ossetian authorities protested the anticrime operation.[16] In late December 2002 and early January 2003, South Ossetian military forces

---

[11] Hockstader, 1994a; Hockstader, 1994b.

[12] Hockstader, 1994a.

[13] "Russian Ship Flees Odessa," *Washington Post*, April 10, 1994.

[14] Boudreaux, 1994, p. 9; Yurii Dubinin, "Bitva za Chernomorskii flot" ["Battle of the Black Sea Fleet'], Rossiya v global'noi politike, No. 1, 2006; Hockstader, 1994b; and "Night Commando Raid Worsens Ukraine-Russia Rift over Fleet," *Los Angeles Times*, April 12, 1994.

[15] Boudreaux, 1994, p. 9; Dubinin, 2006; Felgenhaur, 1999, pp. 11–13; and Kryukov, 2006.

[16] Tea Gularidze, "Concerns Raised over Possible Flare-Up of Violence in Tskhinvali," *Civil Georgia*, February 5, 2003.

and police began mobilizing in Tskhinvali—the capital of the breakaway region—allegedly with Russian materiel support.[17] Russian officials denied bringing weapons and moving heavy equipment into South Ossetia.[18]

This second escalation in tensions resulted in a militarized dispute on February 3, 2003, when the Georgian Ministry of Foreign Affairs stated that the movement of military equipment into South Ossetia by Russia violated the Treaty on Conventional Armed Forces in Europe and requested that Russia remove the weapons from the region.[19] The Georgian presidential envoy to South Ossetia, Vakhtang Rcheulishvili, stated on February 4 that the mobilization of Russian heavy equipment began after Anatoli Chekhov, a Russian parliamentarian, claimed that Georgia intended "to take over the breakaway region with force." "Such statements might further escalate tensions in the regions" responded Rcheulishvili. "Georgia supports only the peaceful settlement of the conflict."[20]

On March 12, 2003, Rcheulishvili traveled to Moscow to discuss a resolution to the conflict.[21] Although public information on the substance of the meeting is not available, it is assumed that the parties discussed the issue of Russian military equipment in South Ossetia. Ultimately, it does not appear that Russia removed any of the equipment despite Georgia's protest.

---

[17] "Russia Refuses Deploying Weapons in Georgia's Breakaway Region," *Civil Georgia*, February 7, 2003; Gularidze, 2003; "Breakaway South Ossetian Forces on High Alert," *Civil Georgia*, January 31, 2003; "Georgia Vice-Speaker to Discuss Abkhaz, South Ossetian Conflicts in Moscow," *Civil Georgia*, March 12, 2003.

[18] Gularidze, 2003.

[19] "Georgia Accuses Russia of Deploying Weapons in South Ossetia," *Civil Georgia*, February 4, 2003.

[20] Gularidze, 2003.

[21] "Georgia Vice-Speaker to Discuss Abkhaz, South Ossetian Conflicts in Moscow," 2003.

Dispute 42: 2003 Abduction of Russian Peacekeeper

On September 27, 2003, a Russian peacekeeping soldier posted in Abkhazia was abducted by four armed men in the town of Zugdidi.[22] In the absence of signs of violence or of a struggle, Zugdidi police officials claimed that the soldier had deserted; the Russian military dismissed this claim.[23] Russian officials believed that the soldier had been kidnapped and would be used as leverage.[24] Subsequently, the kidnappers of the Russian peacekeeper demanded a ransom for the return of the soldier. On September 30, approximately 200 Russian troops conducted searches in Zugdidi in an attempt to find the kidnapped peacekeeper. The Georgian Deputy Foreign Minister, Merab Antadze, stated that Georgia would file a complaint with the CIS peacekeeping headquarters because the Russian patrols in Zugdidi violated the CIS peacekeeping mandate. Georgia contended that Russian troops, posted in Abkhazia under a CIS peacekeeping mandate, were acting unilaterally to search for the kidnapped Russian peacekeeper.[25] On October 1, 2003, as a result of the raids undertaken by the Russian peacekeeping forces, the Russian peacekeeper was released.

---

[22] According to the sources cited in Gibler's report, the date of the abduction was September 27, 2003—not September 20, as stated in the MID description for this dispute (Douglas M. Gibler, *International Conflicts, 1816–2010: Militarized Interstate Dispute Narratives*, Vol. 1, Lanham, Md.: Rowan & Littlefield, 2018, pp. 389–390). Russian peacekeeping forces had been deployed since early 1994 in the breakaway region to act as a buffer. Georgia had consented to their deployment ("Russian Peacekeeper Kidnapped in Georgia Freed," Agence France Presse, October 1, 2003).

[23] Misha Dzhindhikhashvili, "Search Underway in Georgia for Missing Russian Soldier," Associated Press International, September 28, 2003; "Russian Peacekeeper Kidnapped in Georgia: Official," Agence France Presse, September 28, 2003; and "Russian Peacekeeper Kidnapped in Georgia Freed," 2003.

[24] Dzhindhikhashvili, 2003.

[25] "Russian Peacekeepers Search for Kidnapped Solider," *Civil Georgia*, September 30, 2003.

Dispute 43: 2004 Increase in Tensions Between Russia and Georgia

Georgia claimed that Russia violated Georgian airspace several times during the summer and fall of 2004. Separate alleged airspace violations occurred on July 11, August 7, September 14, September 19, October 28, November 13, and November 15.[26] Following the July 11 airspace incident, the Georgian Foreign Ministry claimed that Russia was illegally delivering military equipment to Georgia, the same way (Tbilisi maintained) that Russia had done previously in South Ossetia, and called on Russia to cease shipping arms and military equipment to Georgia.[27] Georgian officials claimed that 160–170 Russian vehicles were moved into the breakaway region of South Ossetia, but the OSCE's mission in Georgia did not find any evidence of this action occurring.[28] Then–President of Georgia Saakashvili claimed that Russia was planning a large-scale armed conflict in South Ossetia.[29] Another incident in September 2004 involved a violation by the Russian military in the Kodori Gorge in the breakaway region of Abkhazia, Georgia.[30] Several other airspace violations during this time were linked with tensions between Russia and Georgia over Chechen terrorists allegedly taking refuge in Georgia.[31]

---

[26] "Georgia Protests Against Violation of Air Space by Russia," *Civil Georgia*, July 12, 2004; "Georgia Protests Violation of Air Space by Russia," *Civil Georgia*, August 7, 2004; "Saakashvili: Certain Forces in Russia Prepare for Aggression Against Georgia," *Civil Georgia*, July 11, 2004; "Controversial Reports over Violation of Georgian Airspace," *Civil Georgia*, September 14, 2004; "Georgian Airspace Violated," *Civil Georgia*, September 19, 2004; "Two Helicopters Violate Georgian Airspace," *Civil Georgia*, October 28, 2004; "Russia Denies Violating Georgia's Airspace," *Civil Georgia*, November 11, 2004; and "Reports: Georgian Airspace Violated Again," *Civil Georgia*, November 16, 2004.

[27] "Georgia Protests Against Violation of Air Space by Russia," 2004.

[28] Giorgi Sepashvili, "Controversial Reports over Deployment of Extra Arms Flare Tensions in Ossetia," *Civil Georgia*, June 13, 2004.

[29] "Saakashvili: Certain Forces in Russia Prepare for Aggression Against Georgia," 2004.

[30] "Georgian Airspace Violated," 2004.

[31] "Georgian Airspace Violated," 2004.

Dispute 44: March 2005–September 2006 Simmering Tensions

On March 22, 2005, 40 Russian peacekeeping troops with seven infantry combat vehicles demanded that Georgian Interior Ministry troops in the village of Ganmukhuri, Abkhazia, give up their weapons.[32] Russian troops withdrew from the village after their leaders spoke with Georgian officials. This incident occurred on the eve of Russia-Georgia negotiations over the conflicts in South Ossetia and Abkhazia. Georgian Foreign Minister Salome Zourabichvili stated that the incident could be a deliberate provocation of the situation by Russia.[33]

In June 2005, the de facto leader of South Ossetia, Eduard Kokoity, stated that Georgia planned to force Russian peacekeeping troops to withdraw from the region.[34] On August 12, Georgian police seized Russian peacekeeping vehicles traveling to Abkhazia and South Ossetia. The main cargo of both vehicles was boxes of cigarettes, which Georgian officials claimed were smuggled.[35] The vehicle that was apprehended in Abkhazia contained 13,000 packs of cigarettes and dozens of bottles of alcohol, although the vehicle that was apprehended in South Ossetia contained 12 boxes of cigarettes and 2 tons of diesel fuel. The vehicle traveling to Abkhazia was seized near the Enguri bridge, which links the region with the rest of Georgia. After the incident, additional Russian peacekeepers and two combat infantry vehicles were deployed to intimidate Georgian forces.[36] The vehicles and their cargo were released on August 15. Forces in both Abkhazia and

---

[32] "Tbilisi Wants Moscow to Explain Incident in Abkhaz Conflict Zone," *Civil Georgia*, March 22, 2005.

[33] "Zourabichvili: Ganmukhuri Incident Could Be a Provocation," *Civil Georgia*, March 22, 2005.

[34] "Troops March in Tskinvali Marking 'Independence,'" *Civil Georgia*, September 20, 2005.

[35] "Reports: Georgia Returns Seized Cargo to Russian Peacekeepers," *Civil Georgia*, August 15, 2005.

[36] "Russian Peacekeepers Accused of Smuggling in Abkhazia, S. Ossetia," *Civil Georgia*, August 12, 2005.

South Ossetia denied that the goods were intended for Russian peacekeeping forces in the area.[37]

On September 20, 2005, mortars fired from the direction of Tbilisi-administered territory targeted Tskhinvali.[38] The attack occurred during a military parade recognizing the 15th anniversary of South Ossetia's secession from Georgia. Georgia had previously protested the military parade, stating that the deployment of troops in Tskhinvali violated the agreements on demilitarization of South Ossetia.[39] The Russian Foreign Ministry published a statement accusing Georgia of bombing Tskhinvali, which immediately followed a resolution by the Georgian Parliament to remove Russian peacekeepers from both South Ossetia and Abkhazia. Following the shelling, a special envoy from Russia, Valery Kenyaikin, traveled between Tbilisi and Tskhinvali to help defuse the situation.[40]

In October 2005, the Georgian Parliament drafted a resolution calling for the withdrawal of Russian peacekeeping troops from South Ossetia by February 2006 and Abkhazia by July 2006. The Russian Foreign Ministry described the draft resolution as provocative.[41] Following the murder of an ethnic Georgian who refused to join Abkhazian forces in November, the Georgian Foreign Ministry stated that the Russian peacekeeping forces failed in their mandate.[42] In January 2006, Georgian Minister of Defense Irakli Okruashvili reiterated the claim that Russian peacekeepers failed to maintain peace in the breakaway regions and that their mandate would not be extended in Georgia.[43]

In April 2006, Russia reopened the Roki Tunnel border crossing between Russia and South Ossetia and the Gantiadi-Adler border crossing between

---

[37] "Reports: Georgia Returns Seized Cargo to Russian Peacekeepers," 2005.

[38] "Russia Slams Georgia over South Ossetia," *Civil Georgia*, October 3, 2005.

[39] "Troops March in Tskinvali Marking 'Independence,'" 2005.

[40] "Russia Slams Georgia over South Ossetia," 2005.

[41] "Georgian-Russian Relations Hit New Low," *Civil Georgia*, October 10, 2005.

[42] "Georgian MFA Condemns Death of Civilian in Gali," *Civil Georgia*, November 6, 2005.

[43] "Defense Minister Speaks Out Against Russian Peacekeepers," *Civil Georgia*, January 5, 2006.

Russia and Abkhazia. Georgia responded by stating that the only legal border crossing between Russia and Georgia was located between Verkhnii Lars (in Russia) and Kazbegi (in Georgia).[44] Russia proceeded to shut down the Verkhnii Lars—Kazbegi border crossing point on July 8, 2006, citing the need for repairs at the station.[45] Georgia increased security measures in the area surrounding the Roki Tunnel.

On September 29, 2006, four Russian officers were sentenced to a two-month pretrial detention following their arrest for spying on Georgia. Three more officers were sentenced to pretrial detention in absentia after refusing to appear in court. Ten Georgian citizens were also arrested and charged with spying for Russia, five of whom confessed to the charges. In response, the Russian Ministry of Emergency Affairs evacuated 100 Russian citizens, including the Russian ambassador to Georgia, from Tbilisi. In Moscow, protests took place outside the Georgian Embassy, with one group of protesters throwing rocks at the Embassy and shattering its windows.[46]

In response to the arrests, the Chairman of the Russian upper house of parliament, Sergei Mironov, stated that Tbilisi was gripped with "spy mania," which could incite a conflict in the regions of Abkhazia and South Ossetia.[47] Russia evacuated its diplomatic staff from its embassy in Tbilisi but suspended efforts to withdraw peacekeeping troops from South Ossetia and Abkhazia.[48] Saakashvili promised to remove any remaining spies located in Georgia, stating that the networks remained mostly in Abkhazia and South Ossetia. Saakashvili also denied that Georgia was mobilizing troops located near Abkhazia.[49]

---

[44] "MFA Slams Russia for Easing Abkhaz Border Crossing," *Civil Georgia*, April 4, 2006.

[45] "Russian Spy Suspects Sentenced to Pre-Trial Detention," *Civil Georgia*, September 29, 2006.

[46] "Russian Spy Suspects Sentenced to Pre-Trial Detention," 2006.

[47] "Senior Russian Senator Speaks of Georgia's 'Spy Mania,'" *Civil Georgia*, September 29, 2006.

[48] "Moscow Tries to Hit Back, as Spy Row Continues," *Civil Georgia*, September 30, 2006.

[49] "Saakashvili Vows to Hunt Down 'Spies' in Abkhazia, S. Ossetia," *Civil Georgia*, October 2, 2006.

Coding Justification for PMC Cases

Capable Adversary's Forces Threatened?

A capable Russian adversary's state forces operated in both Deir Ezzor and Tripoli. Approximately 40 U.S. service members were on the ground in the former case, and Turkish state forces were present in the latter, advising and assisting anti-Hifter forces, operating UAVs, and manning air defense systems.[1] Prior to Hifter's offensive on Tripoli, a small contingent of U.S. forces had been present in the city performing limited advise, assist, and enable activities and conducting counterterrorism operations. However, U.S. forces withdrew from Tripoli days after the beginning of Hifter's offensive.[2]

No capable adversary had official forces present in either the CAR or Cabo Delgado cases. France previously maintained a peacekeeping force in the CAR, its former colony, but had withdrawn it in October 2017 as part of France's general disengagement from the country.[3] The EU Military Training Mission and the UN Multidimensional Integrated Stabilization Mission in the Central African Republic (MINUSCA) are both training the FACA, but they are not adversarial to the Russian presence. On the contrary, Russian PMC trainers at times accompany MINUSCA personnel.[4] The U.S. military undertook humanitarian aid and disaster relief operations

---

[1] Gibbons-Neff, 2018; Ben Fishman and Conor Hiney, "What Turned the Battle for Tripoli?" Washington Institute for Near East Policy, May 6, 2020.

[2] U.S. Africa Command Public Affairs, "Declining Security in Libya Results in Personnel Relocation, Agility Emphasis," press release, Stuttgart, Germany, April 7, 2019.

[3] Searcey, 2019.

[4] Esmenjaud et al., 2018, pp. 7–8.

in Mozambique through April 2019, deploying approximately 100 service members and conducting over 120 sorties following Cyclone Idai.[5] However, the United States does not maintain a large military deployment in the country, nor do other states adversarial to Russia have a large military footprint in Mozambique.

PMCs Tactically Engaged?

Russian PMCs were tactically engaged in Deir Ezzor, Tripoli, and Cabo Delgado. During the clashes in Deir Ezzor, approximately 500 fighters (a mixture of Wagner operatives and regime-aligned Syria militiamen) engaged U.S. forces and their SDF partners. The PMCs and Syrian fighters deployed tanks and other armored vehicles, artillery and mortars, and small arms fire and suffered upward of 200 killed in action.[6] In Libya, Wagner fighters were tactically engaged in the battle for Tripoli as early as September 2019, serving as forward operators and snipers and operating antiair platforms.[7] Approximately 1,200 Wagner personnel deployed to the Tripoli battlespace in late 2019, and as many as 35 were killed in a single Turkish air strike in September 2019.[8] Also beginning in September 2019, approximately 200 Wagner personnel deployed to Cabo Delgado to conduct counterinsurgency operations alongside the FADM in Mozambique. Wagner contractors have engaged in multiple firefights with insurgents, suffering several casualties.[9]

In the CAR, rather than conducting tactical operations, the Wagner personnel's primary roles are training the FACA and guarding infrastructure, including natural resources infrastructure, such as mines, and essential ser-

---

[5] Kyle Rempfer, "U.S. Forces Concluding Relief Efforts in Mozambique," *Air Force Times*, April 12, 2019; and Patrick Tucker, "Mozambique Is Emerging as the Next Islamic Extremist Hotspot," *Defense One*, July 6, 2020.

[6] Gibbons-Neff, 2018; Marten, 2018.

[7] Snow, 2019.

[8] Nichols, 2020; Liliya Yapparova, "A Small Price to Pay for Tripoli," *Meduza*, October 2, 2019.

[9] Sergey Sukhankin, "Russian Mercenaries Pour into Africa and Suffer More Losses (Part One)," *Eurasia Daily Monitor*, Vol. 17, No. 6, January 21, 2020b; and Sauer, 2019.

vices, such as hospitals. However, Wagner PMCs might have been involved in clashes in the CAR. Union for Peace in CAR forces attacked FACA elements whom Russian trainers were accompanying, and one Russian trainer was wounded.[10]

Adversary Partner Force Threatened?

In both Deir Ezzor and Tripoli, Wagner tactical operations threatened an adversary's partner force. In Deir Ezzor, the SDF were coordinating with U.S. special operations forces in the vicinity of the Conoco gas facility, which Wagner and proregime fighters attacked. One SDF fighter was wounded in the incident, the only friendly casualty, and other SDF fighters were involved in the clash.[11] As mentioned already, Turkish state forces deployed to Tripoli to support anti-Hifter groups defending the city against Hifter-aligned forces, including Wagner PMCs. In December 2019, Turkey also began deploying thousands of Syrian contractors to Tripoli to fight Hifter's forces.[12]

Conversely, Wagner's activities in the CAR and Cabo Delgado do not directly threaten an adversarial state's partner forces. In the CAR, other international actors with security roles in the country are de facto aligned with Russia; the EU training mission, in collaboration with MINUSCA, provides support to the FACA, which Russia is also supporting through Wagner trainers. Sudan is also indirectly involved in supporting the FACA, hosting some Russian-led training efforts within its borders.[13] The United States conducts a modest amount of security cooperation with the FADM.[14]

---

[10] Esmenjaud et al., 2018, p. 8.

[11] Gibbons-Neff, 2018; and James N. Mattis, "Media Availability with Secretary Mattis," transcript, Washington, D.C.: U.S. Department of Defense, February 8, 2018.

[12] International Crisis Group, 2020.

[13] European Union External Action, "Common Security and Defence Policy: European Union Training Mission in Central African Republic (EUTM-RCA)," Brussels: European Union, September 19, 2019; Esmenjaud et al., 2018, p. 30.

[14] Between 2015 and 2020, the United States provided approximately $4.5 million worth of security cooperation to Mozambique, $2.5 million of which went to interna-

However, neither the United States nor other states adversarial to Russia maintain significant security relations with the FADM or other Mozambican surrogate forces.

Overt or Covert Russian State Military Forces Present in Immediate Area?

Russian state forces were present in the Tripolitania region, in addition to Wagner PMCs. The role of conventional Russian forces in Tripolitania is unacknowledged and murky; however, there has not been any indication that they are engaged in combat in and around Tripoli. Nevertheless, at least one uniformed Russian soldier has been reported killed in Libya.[15] Russian state forces have also deployed to the CAR. In December 2017, Russia reported to the UN that 170 of the 175 trainers deployed to the CAR were private contractors; the other five were Russian military personnel. In the case of Deir Ezzor, despite Russia's robust military involvement in the Syrian conflict writ large, it does not appear that Russian uniformed military personnel were embedded with the Wagner PMCs or Syrian militia elements involved in the clash. Additionally, there is no indication that Russian state forces are or were operating in Cabo Delgado.

---

tional military and education training. Additionally, in 2019, DoD allocated just under $1 million in Section 333 funds—typically used to aid in partners' counterterrorism, border security, and counterproliferation efforts—to support the FADM (Security Assistance Monitor, "Security Assistance Database," data files, undated). Additionally, General Townsend stated in his 2020 U.S. Africa Command posture statement that humanitarian aid to Mozambique "opened the door . . . for future security cooperation" (Townsend, 2020, p. 15).

[15] "First Russian Soldier Reportedly Dies in Libya, Where the Kremlin Says There Are No Russian Troops," *Meduza*, February 14, 2020; David Schenker, "Assistant Secretary for Near Eastern Affairs David Schenker—Special Briefing," transcript, Washington, D.C.: Press Correspondent's Room, November 26, 2019; and Yapparova, 2019.

Resource-Related Financial Incentive?

The possibility of winning concessions in natural resource industries has incentivized Russian PMC activity in Syria, the CAR, and Mozambique. The Syrian regime's General Petroleum Corporation has to award mining and energy sector rights to Russian companies that reclaim energy infrastructure from Syrian opposition forces, equating to 25 percent of the resource proceeds.[16] Therefore, the Wagner PMCs had a direct incentive to retake the gas plant. In the CAR, Russia has leveraged its political and security support for the central government to extract natural resources concessions—specifically, diamonds and gold.[17] As of February 2020, Africa's three largest liquid natural gas projects are all in Mozambique's Cabo Delgado Province, the epicenter of the country's growing jihadi insurgency. In August 2019, Putin and Mozambican President Filipe Nyusi signed agreements for gas and mineral concessions. Senior U.S. military officers have accused Russia of leveraging Wagner's counterinsurgency activities "in the hopes of buying oil and gas concessions."[18]

Resource-related incentives did not directly drive the Wagner Group's involvement in the battle for Tripoli. Rather, Wagner personnel deployed to Libya's capital to support Hifter's bid to conquer the city, perhaps under contract with the Russian state. However, following the collapse of Hifter's offensive in mid-2020, Wagner forces withdrew from western Libya and redeployed to other areas in the country, including the al-Sharara oil field, the largest in Libya.[19] Hifter has used PMCs to guard oil ports, and Wagner

---

[16] Christopher Spearin, "NATO, Russia and Private Military and Security Companies: Looking into a Dark Reflection," *RUSI Journal*, Vol. 163, No. 3, August 8, 2018, p. 67.

[17] Wagner's owner, Prigozhin, is linked to a mining company, Lobaye Invest, active in the CAR (Searcey, 2019).

[18] Tim Lister and Sebastian Shukla, "Russian Mercenaries Fight Shadowy Battle in Gas-Rich Mozambique," *CNN*, November 29, 2019; and Townsend, 2020, p. 16.

[19] <صنع الله: قلقون من تواجد مرتزقة روس في حق الشرارة> [Sanalla: We Are Unsettled by the Presence of Russian Mercenaries in the al-Sharara Field], *Libya al-Ahrar*, June 24, 2020. The state-owned National Oil Corporation in Tripoli conducts all sales of Libyan oil, and revenue is channeled to the Central Bank of Libya, also in Tripoli. Though officially neutral, both institutions are tacitly aligned because of their locations with the Tripoli-based Government of National Accord, which Hifter attempted to

might deploy again to guard Libyan oil facilities, or, as needed, retake them on behalf of Hifter.[20]

Prior Deployment of PMCs in Country?

Prior to the clashes in Deir Ezzor and Tripoli, Russian PMCs had deployed to Syria and Libya. In the former case, Russian PMCs have been active since at least 2013, when 267 personnel of the Slavonic Corps (a contingent of the Russian PMC Moran Security Group and the precursor to Wagner) deployed to Syria to guard regime-held oil infrastructure. However, they were quickly involved in fighting against the Syrian opposition, suffered heavy casualties, and returned to Russia in October 2013.[21] In addition to Slavonic Corps, RSB Group was also active in Syria prior to the clash in question. In Libya, government officials detained a Russian-led group of contractors in 2012, and RSB Group also deployed to Libya in 2017 to conduct demining and sapping operations in the country's east.[22] By contrast, there is no available reporting that suggests significant Russian PMC activity in the CAR or Mozambique, prior to 2017 and 2019, respectively.

---

overthrow in April 2019. Given that any potential revenue generated from the al-Sharara field would go to the Central Bank and not to Hifter or the eastern-based Libyan government, Wagner likely deployed to al-Sharara to ensure that the field remained under Hifter's control and could be used as leverage in future political negotiations.

[20] "Les Émirats et le Bouclier Noir: Quand des Centaines de Soudanais Sont Envoyés sur le Front Libyen," *Le Vif*, April 29, 2020; "Sudan Investigating Transfer of Guards from UAE to Libyan Oil Port—Ministry," Reuters, January 28, 2020; and Yapparova, 2019.

[21] Rondeaux, 2019, p. 45.

[22] Grzegorz Kuczyński, *Civil War in Libya: Russian Goals and Policy*, Warsaw: Warsaw Institute, April 30, 2019, p. 6; and Sergey Sukhankin, "Continuation of Policy by Other Means: Russian Private Military Contractors in the Libyan Civil War," *Terrorism Monitor*, Vol. 18, No. 3, February 7, 2020c.

Abbreviations

AA	Association Agreement
AOR	area of responsibility
ATO	Anti-Terrorist Operation
CAR	Central African Republic
CINC	Composite Index of National Capability
CIS	Commonwealth of Independent States
DoD	U.S. Department of Defense
EA	electromagnetic attack
ECM	electronic countermeasure
eFP	Enhanced Forward Presence
EU	European Union
EW	electronic warfare
FACA	Central African Armed Forces
FADM	Mozambique Armed Defense Forces
GDP	gross domestic product
GPI	Global Power Index
GPS	global positioning system
IMF	International Monetary Fund
ISIS	Islamic State of Iraq and Syria
LAAF	Libyan Arab Armed Forces
MID	militarized interstate dispute
MINUSCA	UN Multidimensional Integrated Stabilization Mission in the Central African Republic
MRD	Motorized Rifle Division
NAC	North Atlantic Council
NATO	North Atlantic Treaty Organization
NKAO	Nagorno-Karabakh Autonomous Oblast
OIE	operations in the information environment

OSCE	Organization for Security and Co-operation in Europe
PMC	private military company or contractor
SDF	Syrian Democratic Forces
SSR	Soviet Socialist Republic
UAV	unmanned aerial vehicle
UK	United Kingdom
UN	United Nations
USAREUR	U.S. Army Europe

References

Abdullaev, Kamoludin, and Catherine Barnes, eds., *Accord*: Vol. 10, *Politics of Compromise: The Tajikistan Peace Process*, London: Conciliation Resources, 2001.

Akiner, Shirin, and Catherine Barnes, "The Tajik Civil War: Causes and Dynamics," in Abdullaev and Barnes, 2001, pp. 16–22.

Alkhalisi, Zahraa, "Security Experts: Iran-Backed Hackers Targeting U.S. and Saudi Arabia," *CNN Business*, September 21, 2017.

Allison, Roy, "Russia Resurgent? Moscow's Campaign to 'Coerce Georgia to Peace,'" *International Affairs*, Vol. 84, No. 6, November 2008, pp. 1145–1171.

Ambrosio, Thomas, "Insulating Russia from a Color Revolution: How the Kremlin Resists Regional Democratic Trends," *Democratization*, Vol. 14, No. 2, 2007, pp. 232–252.

Antonova, Maria, "They Came to Fight for Ukraine. Now They're Stuck in No Man's Land," *Foreign Policy*, October 19, 2015.

Artman, Vincent M., "Documenting Territory: Passportisation, Territory, and Exception in Abkhazia and South Ossetia," *Geopolitics*, Vol. 18, No. 3, 2013, pp. 682–704.

Asmus, Ronald D., *A Little War That Shook the World*, New York: Palgrave Macmillan, 2010.

Åtland, Kristian, and Torbjørn Pedersen, "The Svalbard Archipelago in Russian Security Policy: Overcoming the Legacy of Fear—or Reproducing It?" *European Security*, Vol. 17, No. 2–3, 2008, pp. 227–251.

Bailey, Michael A., Anton Strezhnev, and Erik Voeten, "Estimating Dynamic State Preferences from United Nations Voting Data," *Journal of Conflict Resolution*, Vol. 61, No. 2, 2017, pp. 430–456.

Balmforth, Richard, and Lina Kushch, "Pro-Moscow Protesters Seize Arms, Declare Republic, Kiev Fears Invasion," Reuters, April 7, 2014.

Barabanov, Mikhail, ed., *Russia's New Army*, Moscow: Centre for Analysis of Strategies and Technologies, 2011.

Baranets, Viktor, "Nachal'nik Genshtaba Vooruzhennykh sil Rosssii General Armii Valerii Gerasimov: 'My Perelomili Khrebet Udarnym Silam Terrorizma,'" *Komsomol'skaya Pravda*, December 26, 2017.

"Belarus Arrests Dozens of Russian Mercenaries: State Media," *Al Jazeera*, July 29, 2020.

Bennett, D. Scott, and Allan C. Stam, "Research Design and Estimator Choices in the Analysis of Interstate Dyads: When Decisions Matter," *Journal of Conflict Resolution*, Vol. 44, No. 5, 2000, pp. 653–685.

Benson, Brett V., "Unpacking Alliances: Deterrent and Compellent Alliances and Their Relationship with Conflict, 1816–2000," *Journal of Politics*, Vol. 73, No. 4, October 2011, pp. 1111–1127.

Biden, Joseph, "Remarks by Vice President Biden in Ukraine," speech given in Kyiv, Ukraine, July 22, 2009.

Black, Ian, and Chris McGreal, "Syria: U.S. Accuses Russia of Sending Attack Helicopters," *The Guardian*, June 12, 2012.

Boudreaux, Richard, "Kiev-Moscow Pact Gives Russia Most of Black Sea Fleet: Accord: Agreement Defuses Volatile Situation. Ukraine Was Approaching a Showdown over the Disputed Armada," *Los Angeles Times*, April 16, 1994.

Braithwaite, Alex, "The Geographic Spread of Militarized Disputes," *Journal of Peace Research*, Vol. 43, No. 5, 2006, pp. 507–522.

Braithwaite, Alex, and Douglas Lemke, "Unpacking Escalation," *Conflict Management and Peace Science*, Vol. 28, No. 2, 2011, pp. 111–123.

Braw, Elisabeth, "The GPS Wars Are Here," *Foreign Policy*, December 17, 2018.

"Breakaway South Ossetian Forces on High Alert," *Civil Georgia*, January 31, 2003.

Brown, Bess A., "The Civil War in Tajikistan, 1992–1993," in Mohammad-Reza Djalili, Frédéric Grare, and Shirin Akiner, eds., *Tajikistan: The Trials of Independence*, 1st ed., New York: Routledge, 1998, pp. 86–96.

Browne, Ryan, "Top U.S. General Warns Russia Using Mercenaries to Access Africa's Natural Resources," *CNN*, February 7, 2019.

Brudenell, Anna Maria, "Russia's Role in the Kosovo Conflict of 1999," *RUSI Journal*, Vol. 153, No. 1, 2008, pp. 30–34.

Buchanan, Ben, "Five Myths About Cyberwar," *Washington Post*, February 20, 2020.

Bueno de Mesquita, Bruce, James D. Morrow, and Ethan R. Zorick, "Capabilities, Perception, and Escalation," *The American Political Science Review*, Vol. 91, No. 1, March 1997, pp. 15–27.

Carlson, Lisa J., "A Theory of Escalation and International Conflict," *Journal of Conflict Resolution*, Vol. 39, No. 3, September 1995, pp. 511–534.

Carter, David B., and Curtis S. Signorino, "Back to the Future: Modeling Time Dependence in Binary Data," *Political Analysis*, Vol. 18, No. 3, 2010, pp. 271–292.

Caspersen, Nina, and Gareth Stansfield, eds., *Unrecognized States in the International System*, New York: Routledge, 2011.

Charap, Samuel, and Timothy J. Colton, *Everyone Loses: The Ukraine Crisis and the Ruinous Contest for Post-Soviet Eurasia*, New York: Routledge, January 2017.

Charap, Samuel, Elina Treyger, and Edward Geist, *Understanding Russia's Intervention in Syria*, Santa Monica, Calif.: RAND Corporation, RR-3180-AF, 2019. As of October 31, 2020:
https://www.rand.org/pubs/research_reports/RR3180.html

Chebankova, Elena, "Russia's Idea of the Multipolar World Order: Origins and Main Dimensions," *Post-Soviet Affairs*, Vol. 33, No. 3, 2017, pp. 217–234.

Clark, Susan L., *The Central Asian States: Defining Security Priorities and Developing Military Forces*, Alexandria, Va.: Institute for Defense Analyses, IDA Paper P-2886, September 1993.

"Clashes in Eastern Ukraine Reportedly Turn Deadly," NPR, April 13, 2014.

Connable, Ben, Stephanie Young, Stephanie Pezard, Andrew Radin, Raphael S. Cohen, Katya Migacheva, and James Sladden, *Russia's Hostile Measures: Combating Russian Gray Zone Aggression Against NATO in the Contact, Blunt, and Surge Layers of Competition*, Santa Monica, Calf.: RAND Corporation, RR-2539-A, 2020. As of July 28, 2020:
https://www.rand.org/pubs/research_reports/RR2539.html

"Controversial Reports over Violation of Georgian Airspace," *Civil Georgia*, September 14, 2004.

Correlates of War Project, "State System Membership List," ver. 2016, 2017. As of October 1, 2020:
http://correlatesofwar.org

Cotton, Sarah K., Ulrich Petersohn, Molly Dunigan, Q. Burkhart, Megan Zander-Cotugno, Edward O'Connell, and Michael Webber, *Hired Guns: Views About Armed Contractors in Operation Iraqi Freedom*, Santa Monica, Calif.: RAND Corporation, MG-987-SRF, 2010. As of October 1, 2020:
https://www.rand.org/pubs/monographs/MG987.html

D'Anieri, Paul, *Ukraine and Russia: From Civilized Divorce to Uncivil War*, Cambridge, United Kingdom: Cambridge University Press, 2019.

Dawar, Anil, "Putin Warns NATO over Expansion," *The Guardian*, April 4, 2008.

De Waal, Thomas, *The Caucasus: An Introduction*, New York: Oxford University Press, 2010.

"Defense Minister Speaks Out Against Russian Peacekeepers," *Civil Georgia*, January 5, 2006.

Delanoe, Igor, *Russia's Black Sea Fleet: Toward a Multiregional Force*, Arlington, Va.: Center for Naval Analyses, June 2019.

Diehl, Paul F., "Arms Races and Escalation: A Closer Look," *Journal of Peace Research*, Vol. 20, No. 3, September 1983, pp. 205–212.

———, "Contiguity and Military Escalation in Major Power Rivalries, 1816–1980," *Journal of Politics*, Vol. 47, No. 4, November 1985, pp. 1203–1211.

Dixon, William J., "Third-Party Techniques for Preventing Conflict Escalation and Promoting Peaceful Settlement," *International Organization*, Vol. 50, No. 4, Autumn 1996, pp. 653–681.

Djalili, Mohammad-Reza, and Frédéric Grare, "Regional Ambitions and Interests in Tajikistan: The Role of Afghanistan, Pakistan, and Iran," in Mohammad-Reza Djalili, Frédéric Grare, and Shirin Akiner, eds., *Tajikistan: The Trials of Independence*, New York: Routledge, 1998, pp. 119–131.

DoD—*See* U.S. Department of Defense.

Dreyfus, Emmanuel, *Private Military Companies in Russia: Not So Quiet on the Eastern Front?* Paris: Insitut de Recherche Stratégique de L'École Militaire, October 12, 2018. As of August 28, 2020: https://www.irsem.fr/data/files/irsem/documents/document/file/2955/RP_IRSEM_No_63_2018.pdf

Dubinin, Yurii, "Bitva za Chernomorskii Flot [Battle of the Black Sea Fleet]," *Rossiya v global'noi politike*, No. 1, 2006.

Dudoignon, Stéphane A., "Political Parties and Forces in Tajikistan, 1989–1993," in Mohammad-Reza Djalili, Frédéric Grare, and Shirin Akiner, eds., *Tajikistan: The Trials of Independence*, New York: Routledge, 1998, pp. 52–85.

Dzhindhikhashvili, Misha, "Search Underway in Georgia for Missing Russian Soldier," Associated Press International, September 28, 2003.

Eckel, Mike, "New Scrutiny for 'Putin's Chef' and Russian Mercenaries in Africa," Radio Free Europe/Radio Liberty, October 1, 2019.

Elder, Miriam, "Syria Will Receive Attack Helicopters from Russia, Kremlin Confirms," *The Guardian*, June 28, 2012.

El Gomati, Anas, "Russia's Role in the Libyan Civil War Gives It Leverage over Europe," *Foreign Policy*, January 18, 2020.

"Les Émirats et le Bouclier Noir: Quand des Centaines de Soudanais Sont Envoyés sur le Front Libyen]," *Le Vif*, April 29, 2020.

Entous, Adam, and Julian E. Barnes, "Russian Buildup Stokes Worries: Pentagon Alarmed as Troops Mass near Ukraine Border," *Wall Street Journal*, March 28, 2014.

Epkenhans, Tim, *The Origins of the Civil War in Tajikistan: Nationalism, Islamism and Violent Conflict in Post-Soviet Space*, Lanham, Md.: Lexington Books, 2016.

Esmenjaud, Romain, Mélanie De Groof, Paul-Simon Handy, Ilyas Oussedik, and Enrica Picco, *Midterm Report of the Panel of Experts on the Central African Republic Extended Pursuant to Security Council Resolution 2399*, New York: United Nations Security Council, July 23, 2018.

Esmenjaud, Romain, Mélanie De Groof, Ilyas Oussedik, Anna Osborne, and Émile Rwagasana, *Final Report of the Panel of Experts on the Central African Republic Extended Pursuant to Security Council Resolution 2454*, New York: United Nations Security Council, December 14, 2019.

European Union External Action, "Common Security and Defence Policy: European Union Training Mission in Central African Republic (EUTM-RCA)," Brussels: European Union, September 19, 2019.

Fainberg, Sarah, "Russian Spetsnaz: Contractors and Volunteers in the Syrian Conflict," *Russie.NEI.Visions*, No. 105, French Institute of International Relations, December 2017.

Fazzini, Kate, "The Saudi Oil Attacks Could Be a Precursor to Widespread Cyberwarfare—with Collateral Damage for Companies in the Region," *CNBC*, September 21, 2019.

Fearon, James D., "Signaling Versus the Balance of Power and Interests: An Empirical Test of a Crisis Bargaining Model," *Journal of Conflict Resolution*, Vol. 38, No. 2, *Arms, Alliances, and Cooperation: Formal Models and Empirical Tests*, June 1994a, pp. 236–269.

———, "Domestic Political Audiences and the Escalation of International Disputes," *American Political Science Review*, Vol. 88, No. 3, September 1994b, pp. 577–592.

———, "Rationalist Explanations for War," *International Organization*, Vol. 49, No. 3, Summer 1995, pp. 379–414.

Felgenhauer, Tyler, *Ukraine, Russia, and the Black Sea Fleet Accords*, Princeton, N.J.: Woodrow Wilson School of Public and International Affairs, WWS Case Study 2/99, February 26, 1999.

"First Russian Soldier Reportedly Dies in Libya, Where the Kremlin Says There Are No Russian Troops," *Meduza*, February 14, 2020.

Firth, David, "Bias Reduction of Maximum Likelihood Estimates," *Biometrika*, Vol. 80, No. 1, March 1993, pp. 27–38.

Fishman, Ben, and Conor Hiney, "What Turned the Battle for Tripoli?" Washington Institute for Near East Policy, May 6, 2020.

Flanagan, Jane, "Mozambique Calls on Russian Firepower," *The Times* (London), October 2, 2019.

Fravel, M. Taylor, "Power Shifts and Escalation: Explaining China's Use of Force in Territorial Disputes," *International Security*, Vol. 32, No. 3, Winter 2007/08, pp. 44–83.

Frizell, Sam, "Violence in East Ukraine Ratchets Up Tensions with Russia," *TIME*, April 6, 2014.

Gardner, Frank, "How Vital Is Syria's Tartus Port to Russia?" *BBC News*, June 27, 2012.

Geller, Daniel S., "Power Differentials and War in Rival Dyads," *International Studies Quarterly*, Vol. 37, No. 2, June 1993, pp. 173–193.

"Georgia Accuses Russia of Deploying Weapons in South Ossetia," *Civil Georgia*, February 4, 2003.

"Georgia Protests Against Violation of Air Space by Russia," *Civil Georgia*, July 12, 2004.

"Georgia Protests Violation of Air Space by Russia," *Civil Georgia*, August 7, 2004

"Georgia Vice-Speaker to Discuss Abkhaz, South Ossetian Conflicts in Moscow," *Civil Georgia*, March 12, 2003.

"Georgian Airspace Violated," *Civil Georgia*, September 19, 2004.

"Georgian MFA Condemns Death of Civilian in Gali," *Civil Georgia*, November 6, 2005.

"Georgian-Russian Relations Hit New Low," *Civil Georgia*, October 10, 2005.

Ghatak, Sambuddha, Aaron Gold, and Brandon C. Prins, "External Threat and the Limits of Democratic Pacifism," *Conflict Management and Peace Science*, Vol. 34, No. 2, March 2017, pp. 141–159.

Gibbons-Neff, Thomas, "How a 4-Hour Battle Between Russian Mercenaries and U.S. Commandos Unfolded in Syria," *New York Times*, May 24, 2018.

Gibler, Douglas M., *International Conflicts, 1816–2010: Militarized Interstate Dispute Narratives*, Vol. 1, Lanham, Md.: Rowan & Littlefield, 2018.

Gibler, Douglas M., Steven V. Miller, and Erin K. Little, "An Analysis of the Militarized Interstate Dispute (MID) Dataset, 1816–2001," *International Studies Quarterly*, Vol. 60, No. 4, December 2016, pp. 719–730.

Giles, Keir, *Assessing Russia's Reorganized and Rearmed Military*, Washington, D.C.: Carnegie Endowment for International Peace, May 3, 2017.

Glaser, Charles L., "Political Consequences of Military Strategy: Expanding and Refining the Spiral and Deterrence Models," *World Politics*, Vol. 44, No. 4, July 1992, pp. 497–538.

———, "Realists as Optimists: Cooperation as Self-Help," in Michael E. Brown, Owen R. Coté, Jr., Sean M. Lynn-Jones, and Steven E. Miller, eds., *Theories of War and Peace: An International Security Reader*, Cambridge, Mass.: MIT Press, 1998, pp. 94–135.

Glazunova, Lyubov, "Russia Is Washing Blood Off African Diamonds," *RIDDLE*, June 9, 2020.

Gleason, Gregory, "Why Russia Is in Tajikistan," *Comparative Strategy*, Vol. 20, No. 1, 2001, pp. 77–89.

Gleditsch, Nils Petter, and Håvard Hegre, "Peace and Democracy: Three Levels of Analysis," *Journal of Conflict Resolution*, Vol. 41, No. 2, April 1997, pp. 283–310.

Gochman, Charles S., and Russell J. Leng, "Realpolitik and the Road to War: An Analysis of Attributes and Behavior," *International Studies Quarterly*, Vol. 27, No. 1, March 1983, pp. 97–120.

Götz, Elias, "It's Geopolitics, Stupid: Explaining Russia's Ukraine Policy," *Global Affairs*, Vol. 1, No. 1, 2015, pp. 3–10.

Government of the Russian Federation, *Soglashenie mezhdu Rossiiskoi Federatsiei i Ukrainoi o printsipakh formirovaniya VMF Rossii i VMS Ukrainy na baze Chernomorskogo flota byvshego SSSR*, Moscow, August 3, 1992.

———, *Protokol ob uregulirovaniya problem Chernomorskogo flota*, Moscow, September 3, 1993.

Greenberg, Andy, "The Highly Dangerous 'Triton' Hackers Have Probed the U.S. Grid," *Wired*, June 14, 2019.

Grytsenko, Oksana, "Thousands of Russian Soldiers Fought at Ilovaisk, Around a Hundred Were Killed," *Kyiv Post*, April 6, 2018.

Gularidze, Tea, "Concerns Raised over Possible Flare-Up of Violence in Tskhinvali," *Civil Georgia*, February 5, 2003.

Gurses, Mehmet, "Transnational Ethnic Kin and Civil War Outcomes," *Political Research Quarterly*, Vol. 68, No. 1, March 2015, pp. 142–153.

Hauer, Neil, "Putin Has a New Secret Weapon in Syria: Chechens," *Foreign Policy*, May 4, 2017.

Heim, Jacob L., and Benjamin M. Miller, *Measuring Power, Power Cycles, and the Risk of Great-Power War in the 21st Century*, Santa Monica, Calif.: RAND Corporation, RR-2989-RC, 2020. As of October 31, 2020:
https://www.rand.org/pubs/research_reports/RR2989.html

Hensel, Paul R., "Charting a Course to Conflict: Territorial Issues and Interstate Conflict, 1816–1992," *Conflict Management and Peace Science*, Vol. 15, No. 1, 1996, pp. 43–73.

———, "Contentious Issues and World Politics: The Management of Territorial Claims in the Americas, 1816–1992," *International Studies Quarterly*, Vol. 45, No. 1, March 2001, pp. 81–109.

Hensel, Paul R., and Bryan Frederick, "Provisional Issue Correlates of War Territorial Claims Dataset," version 1.02, unpublished manuscript, January 1, 2017.

Hensel, Paul R., Sara McLaughlin Mitchell, and Thomas E. Sowers II, "Conflict Management of Riparian Disputes," *Political Geography*, Vol. 25, No. 4, May 2006, pp. 383–411.

Hensel, Paul R., Sara McLaughlin Mitchell, Thomas E. Sowers II, and Clayton L. Thyne, "Bones of Contention: Comparing Territorial, Maritime, and River Issues," *Journal of Conflict Resolution*, Vol. 52, No. 1, February 2008, pp. 117–143.

Herszenhorn, David M., Patrick Reevell, and Noah Sneider, "Russian Forces Take Over One of the Last Ukrainian Bases in Crimea," *New York Times*, March 22, 2014.

Herszenhorn, David M., and Andrew Roth, "In East Ukraine, Protesters Seek Russian Troops," *New York Times*, April 7, 2014.

Hill, Fiona, "Mr. Putin and the Art of the Offensive Defense: Approaches to Foreign Policy (Part Two)," Brookings Institution, March 16, 2014.

———, "Putin Battles for the Russian Homefront in Syria," Brookings Institution, May 23, 2016.

Hill, William H., *No Place for Russia: European Security Institutions Since 1989*, New York: Columbia University Press, 2018.

Hockstader, Lee, "Brush with Black Sea Naval Battle Heightens Russo-Ukrainian Tensions; Warships, Fighter Jets Dispatched in Weekend Confrontation," *Washington Post*, April 11, 1994a.

———, "Ukraine Detains Officers After Russia Grabs Ship, as Fleet Conflict Escalates," *Washington Post*, April 12, 1994b.

Hodgson, Quentin E., Logan Ma, Krystyna Marcinek, and Karen Schwindt, *Fighting Shadows in the Dark: Understanding and Countering Coercion in Cyberspace*, Santa Monica, Calif.: RAND Corporation, RR-2961-OSD, 2019. As of August 17, 2021:
https://www.rand.org/pubs/research_reports/RR2961.html

Howell, William G., and Jon C. Pevehouse, *While Dangers Gather: Congressional Checks on Presidential War Powers*, Princeton, N.J.: Princeton University Press, 2007.

Human Rights Watch, *Singled Out: Russia's Detention and Expulsion of Georgians*, Vol. 19, No. 5(D), New York, October 2007. As of September 1, 2020:
http://www.hrw.org/sites/default/files/reports/russia1007webwcover.pdf

Huth, Paul K., and Todd L. Allee, *The Democratic Peace and Territorial Conflict in the Twentieth Century*, Cambridge, England: Cambridge University Press, 2003.

Huth, Paul, Christopher Gelpi, and D. Scott Bennett, "The Escalation of Great Power Militarized Disputes: Testing Rational Deterrence Theory and Structural Realism," *American Political Science Review*, Vol. 87, No. 3, September 1993, pp. 609–623.

Huth, Paul, and Bruce Russett, "Deterrence Failure and Crisis Escalation," *International Studies Quarterly*, Vol. 32, No. 1, March 1988, pp. 29–45.

Independent International Fact-Finding Mission on the Conflict in Georgia, *Independent International Fact-Finding Mission on the Conflict in Georgia Report*, Vol. I, Brussels: Council of the European Union, September 2009.

International Crisis Group, *Turkey Wades into Libya's Troubled Waters*, Brussels, April 30, 2020.

Jackson, Patrick, "Ukraine War Pulls In Foreign Fighters," *BBC News*, September 1, 2014.

Jervis, Robert, "Cooperation Under the Security Dilemma," *World Politics*, Vol. 30, No. 2, January 1978, pp. 167–214.

Jones, Daniel M., Stuart A. Bremer, and J. David Singer, "Militarized Interstate Disputes, 1816–1992: Rationale, Coding Rules, and Empirical Patterns," *Conflict Management and Peace Science*, Vol. 15, No. 2, September 1996, pp. 163–213.

Kaufman, Stuart J., *Modern Hatreds: The Symbolic Politics of Ethnic War*, Ithaca, N.Y.: Cornell University Press, 2001.

Kay, John, and Mervyn King, *Radical Uncertainty: Decision-Making Beyond the Numbers*, New York: W.W. Norton & Company, 2020.

King, Gary, and Langche Zeng, "Logistic Regression in Rare Events Data," *Political Analysis*, Vol. 9, No. 2, 2001, pp. 137–163.

Kinsella, David, and Bruce Russett, "Conflict Emergence and Escalation in Interactive International Dyads," *Journal of Politics*, Vol. 64, No. 4, November 2002, pp. 1045–1068.

Kofman, Michael, Katya Migacheva, Brian Nichiporuk, Andrew Radin, Olesya Tkacheva, and Jenny Oberholtzer, *Lessons from Russia's Operations in Crimea and Eastern Ukraine*, Santa Monica, Calif.: RAND Corporation, RR-1498-A, 2017. As of October 31, 2020:
https://www.rand.org/pubs/research_reports/RR1498.html

Kolstø, Pål, "The Sustainability and Future of Unrecognized Quasi-States," *Journal of Peace Research*, Vol. 43, No. 6, 2006, pp. 723–740.

Korostelina, Karina, "Shaping Unpredictable Past: National Identity and History Education in Ukraine," *National Identities*, Vol. 13, No. 1, March 2011, pp. 1–16.

Kramer, Andrew E., and Andrew Higgins, "Ukraine Forces Storm a Town, Defying Russia," *New York Times*, April 13, 2014.

Kryukov, N. A., "Osobennosti razvitiya i sostoyaniya rossiisko-ukrainskikh otnoshenii po pravovomu statusu Chernomorskogo flota RF," *Voennaya mysl'*, No. 5, 2006.

Kuczyński, Grzegorz, *Civil War in Libya: Russian Goals and Policy*, Warsaw: Warsaw Institute, April 30, 2019.

Kugler, Jacek, and Douglas Lemke, "The Power Transition Research Program: Assessing Theoretical and Empirical Advances," in Manus I. Midlarsky, ed., *Handbook of War Studies II*, Ann Arbor, Mich.: University of Michigan Press, 2000, pp. 129–163.

"Kuril Islands Dispute Between Russia and Japan," *BBC News*, April 29, 2013.

Kushch, Lina, "Pro-Russia Protesters Occupy Regional Government in Ukraine's Donetsk," Reuters, March 3, 2014.

Kushch, Lina, and Thomas Grove, "Pro-Russia Protesters Seize Ukraine Buildings, Kiev Blames Putin," Reuters, April 6, 2014.

Laitin, David D., "Identity in Formation: The Russian-Speaking Nationality in the Post-Soviet Diaspora," *European Journal of Sociology*, Vol. 36, No. 2, 1995, pp. 281–316.

Lampert, Brian, "Putin's Prospects: Vladimir Putin's Decision-Making Through the Lens of Prospect Theory," *Small Wars Journal*, February 15, 2016.

Lannin, Patrick, "Russia, Latvia Finally Seal Border Treaty," Reuters, December 18, 2007.

Lavrov, A. V., "Khod boevykh deistvii v 2011–2015 godakh," in M. Yu. Shepovalenko, ed., *Siriiskii rubezh*, Moscow: Tsentr analiza strategii i tekhnologii, 2016.

Lavrov, Sergei, "NATO Expansion a Huge Mistake," *Interfax*, December 12, 2006.

————, "Za i PROtiv: Sergei Lavrov o vneshnepoliticheskikh vragakh, o vozmozhnoi voine mezhdu SShA i Iranom i mnogom drugom," *Rossiiskaya Gazeta*, October 24, 2012.

————, "Russia's Foreign Policy Philosophy," *International Affairs*, March 2013.

Lazareva, Inna, "Russian Spy Base in Syria Used to Monitor Rebels and Israel Seized," *The Telegraph*, October 8, 2014.

Leeds, Brett Ashley, "Do Alliances Deter Aggression? The Influence of Military Alliances on the Initiation of Militarized Interstate Disputes," *American Journal of Political Science*, Vol. 47, No. 3, July 2003, pp. 427–439.

Leeds, Brett Ashley, and David R. Davis, "Domestic Political Vulnerability and International Disputes," *Journal of Conflict Resolution*, Vol. 41, No. 6, 1997, pp. 814–834.

Lemke, Douglas, *Regions of War and Peace*, New York: Cambridge University Press, 2002.

Lemke, Douglas, and William Reed, "Regime Types and Status Quo Evaluations: Power Transition Theory and the Democratic Peace," *International Interactions*, Vol. 22, No. 2, 1996, pp. 143–164.

Leng, Russell J., and J. David Singer, "Militarized Interstate Crises: The BCOW Typology and Its Applications," *International Studies Quarterly*, Vol. 32, No. 2, June 1988, pp. 155–173.

Levinsson, Claess, "The Long Shadow of History: Post-Soviet Border Disputes—The Case of Estonia, Latvia, and Russia," *Connections*, Vol. 5, No. 2, Fall 2006, pp. 98–109.

Levy, Jack S., "The Diversionary Theory of War: A Critique," in Manus I. Midlarsky, ed., *Handbook of War Studies*, Boston: Unwin Hyman, 1989, pp. 259–288.

Lewis, Paul, "UN Ambassador: Ukraine Unrest Has 'Tell-Tale Signs of Moscow's Involvement,'" *The Guardian*, April 13, 2014.

Linder, Andrew, *Russian Private Military Companies in Syria and Beyond*, Washington, D.C.: Center for Strategic and International Studies, undated.

Linzer, Dafna, Jeff Larson, and Michael Grabell, "Flight Records Say Russia Sent Syria Tons of Cash," *ProPublica*, November 26, 2012.

Lister, Tim, and Sebastian Shukla, "Russian Mercenaries Fight Shadowy Battle in Gas-Rich Mozambique," *CNN*, November 29, 2019.

Lister, Tim, Sebastian Shukla, and Clarissa Ward, "CNN Special Report: Putin's Private Army," *CNN*, August 2019.

Lynch, Allen C., "The Evolution of Russian Foreign Policy in the 1990s," *Journal of Communist Studies and Transition Politics*, Vol. 18, No. 1, 2002, pp. 161–182.

Lynch, Colum, "Why Putin Is So Committed to Keeping Assad in Power," *Foreign Policy*, October 7, 2015.

Lynch, Dov, *Russian Peacekeeping Strategies in the CIS: The Cases of Moldova, Georgia, and Tajikistan*, London: Royal Institute of International Affairs, 2000.

——, "The Tajik Civil War and Peace Process," *Civil Wars*, Vol. 4, No. 4, 2001, pp. 49–72.

Malyarenko, Tetyana, and David J. Galbreath, "Paramilitary Motivation in Ukraine: Beyond Integration and Abolition," *Southeast European and Black Sea Studies*, Vol. 16, No. 1, 2016, pp. 113–138.

Mankoff, Jeffery, *Russian Foreign Policy: The Return of Great Power Politics*, Lanham, Md.: Rowman & Littlefield, 2009.

Maoz, Zeev, "Resolve, Capabilities, and the Outcomes of Interstate Disputes, 1816–1976," *Journal of Conflict Resolution*, Vol. 27, No. 2, June 1983, pp. 195–229.

Marshall, Monty G., Ted Robert Gurr, and Keith Jaggers, *Polity IV Project: Political Regime Characteristics and Transitions, 1800–2018: Dataset Users' Manual*, Vienna, Va.: Center for Systemic Peace, July 27, 2019.

Marten, Kimberly, "The Puzzle of Russian Behavior in Deir al-Zour," *War on the Rocks*, July 5, 2018.

——, "Russia's Use of Semi-State Security Forces: The Case of the Wagner Group," *Post-Soviet Affairs*, Vol. 35, No. 3, 2019, pp. 181–204.

Mattis, James N., "Media Availability with Secretary Mattis," transcript, Washington, D.C.: U.S. Department of Defense, February 8, 2018.

Matveeva, Anna, *The Perils of Emerging Statehood: Civil War and State Reconstruction in Tajikistan: An Analytical Narrative on State-Making*, London: Crisis States Research Centre, Working Paper No. 46, March 2009.

Maxwell, Neville, "How the Sino-Russian Boundary Conflict Was Finally Settled: From Nerchinsk 1689 to Vladivostok 2005 via Zhenbao Island 1969," in Iwashita Akihiro, ed., *Eager Eyes Fixed on Eurasia*, Vol. 2, *Russia and Its Eastern Edge*, Sopporo, Japan: Slavic Research Center, Hokkaido University, 21st Century COE Program, Slavic Eurasian Studies, No. 16-2, 2007, pp. 229–253.

McDermott, Roger N., *Russia's Electronic Warfare Capabilities to 2025: Challenging NATO in the Electromagnetic Spectrum*, Tallinn, Estonia: International Center for Defence and Security, September 2017.

McGregor, Andrew, "Falling off the Fence: Russian Mercenaries Join the Battle for Tripoli," *Eurasia Daily Monitor*, Vol. 16, No. 138, October 8, 2019.

McLeary, Paul, "Russians Tried to Jam NATO Exercise; Swedes Say They've Seen This Before," *Breaking Defense*, November 20, 2018.

McLees, Alexandra, and Eugene Rumer, "Saving Ukraine's Defense Industry," Carnegie Endowment for International Peace, July 30, 2014.

Melin, Molly M., and Alexandru Grigorescu, "Connecting the Dots: Dispute Resolution and Escalation in a World of Entangled Territorial Claims," *Journal of Conflict Resolution*, Vol. 58, No. 6, 2014, pp. 1085–1109.

Menon, Rajan, "After Empire: Russia and the Southern 'Near Abroad,'" in Michael Mandelbaum, ed., *The New Russian Foreign Policy*, New York: Council on Foreign Relations, 1998.

Menon, Rajan, and Eugene Rumer, *Conflict in Ukraine: The Unwinding of the Post-Cold War Order*, Boston: MIT Press, 2015.

Merry, E. Wayne, "The Origins of Russia's War in Ukraine: The Clash of Russian and European 'Civilizational Choices' for Ukraine," in Elizabeth A. Wood, William E. Pomeranz, E. Wayne Merry, and Maxim Trudolyubov, *Roots of Russia's War in Ukraine*, Washington, D.C.: Woodrow Wilson Center Press, 2015, pp. 27–50.

"MFA Slams Russia for Easing Abkhaz Border Crossing," *Civil Georgia*, April 14, 2006.

Miller, Zeke J., "Obama: U.S. Working to 'Isolate Russia,'" *TIME*, March 3, 2014.

Ministry of Foreign Affairs of the Russian Federation, *The Foreign Policy Concept of the Russian Federation*, translation, Moscow, July 12, 2008.

Minney, Leslie, Rachel Sullivan, and Rachel Vandenbrink, "Amid the Central African Republic's Search for Peace, Russia Steps In. Is China Next?" United States Institute for Peace, December 19, 2019.

Morgan, Forrest E., Karl P. Mueller, Evan S. Medeiros, Kevin L. Pollpeter, and Roger Cliff, *Dangerous Thresholds: Managing Escalation in the 21st Century*, Santa Monica, Calif.: RAND Corporation, MG-614-AF, 2008. As of July 20, 2020:
https://www.rand.org/pubs/monographs/MG614.html

Morgan, T. Clifton, and Kenneth N. Bickers, "Domestic Discontent and the External Use of Force," *Journal of Conflict Resolution*, Vol. 36, No. 1, March 1992, pp. 25–52.

Morrow, James D., "Capabilities, Uncertainty, and Resolve: A Limited Information Model of Crisis Bargaining," *American Journal of Political Science*, Vol. 33, No. 4, November 1989, pp. 941–972.

"Moscow Tries to Hit Back, as Spy Row Continues," *Civil Georgia*, September 30, 2006.

Mostovaya, Yuliya, Sergei Rakhmanin, and Inna Vedernikova, "Yugo-Vostok: vetv' dreva nashego," *Zerkalo nedeli*, April 18, 2014.

Moyer, Jonathan D., and Alanna Markle, *Relative National Power Codebook*, Denver, Colo.: Frederick S. Pardee Center for International Futures, Josef Korbel School of International Studies, University of Denver, version 7.2.2018, 2017.

Mueller, John E., *War, Presidents, and Public Opinion*, New York: John Wiley & Sons, 1973.

Myre, Greg, "The Military Doesn't Advertise It, but U.S. Troops Are All over Africa," NPR, April 28, 2018.

NATO—*See* North Atlantic Treaty Organization.

Nichols, Michelle, "Up to 1,200 Deployed in Libya by Russian Military Group: U.N. Report," Reuters, May 6, 2020.

"Night Commando Raid Worsens Ukraine-Russia Rift over Fleet," *Los Angeles Times*, April 12, 1994.

Niou, Emerson M. S., and Sean M. Zeigler, "External Threat, Internal Rivalry, and Alliance Formation," *Journal of Politics*, Vol. 81, No. 2, April 2019, pp. 571–584.

Norris, John, *Collision Course: NATO, Russia, and Kosovo*, Santa Barbara, Calif.: Greenwood Publishing Group, Praeger Publishers, 2005.

North Atlantic Treaty Organization, "Study on NATO Enlargement," webpage, September 3, 1995. As of September 1, 2020:
https://www.nato.int/cps/en/natohq/official_texts_24733.htm?#top

———, "Bucharest Summit Declaration: Issued by the Heads of State and Government Participating in the Meeting of the North Atlantic Council in Bucharest on 3 April 2008," webpage, May 8, 2014.

Norwegian Intelligence Service, *Focus 2019: The Norwegian Intelligence Service's Assessment of Current Security Challenges*, Oslo, January 21, 2019.

Odom, William E., *The Collapse of the Soviet Military*, New Haven, Conn.: Yale University Press, 2000.

O'Dwyer, Gerard, "Finland to Bolster Navy's Surface Fleet with New Ships, More Missiles," *Defense News*, January 13, 2020.

Oneal, John R., and Bruce Russett, "Rule of Three, Let It Be? When More Really Is Better," *Conflict Management and Peace Science*, Vol. 22, No. 4, September 2005, pp. 293–310.

Organski, A. F. K., and Jacek Kugler, *The War Ledger*, Chicago: University of Chicago Press, 1980.

Orr, Michael, "The Russian Army and the War in Tajikistan (with Map)," in Mohammad-Reza Djalili, Frédéric Grare, and Shirin Akiner, eds., *Tajikistan: The Trials of Independence*, New York: Routledge, 1998, pp. 151–160.

Palmer, Glenn, Vito D'Orazio, Michael R. Kenwick, and Roseanne W. McManus, "Updating the Militarized Interstate Dispute Data: A Response to Gibler, Miller, and Little," *International Studies Quarterly*, Vol. 64, No. 2, June 2020, pp. 469–475.

Parker, John W., *Putin's Syrian Gambit: Sharper Elbows, Bigger Footprint, Stickier Wicket*, Washington, D.C.: Center for Strategic Research, Institute for National Strategic Studies, National Defense University, Strategic Perspectives No. 25, July 2017.

Partell, Peter J., "Escalation at the Outset: An Analysis of Targets' Responses in Militarized Interstate Disputes," *International Interactions*, Vol. 23, No. 1, 1997, pp. 1–35.

Pegg, Scott, *International Society and the De Facto State*, Brookfield, Vt.: Ashgate, 1998.

Perlroth, Nicole, and Clifford Krauss, "A Cyberattack in Saudi Arabia Had a Deadly Goal. Experts Fear Another Try," *New York Times*, March 15, 2018.

Persen, Kjell, "This Is How Norway Was Jammed During the NATO Exercise," *TV2 Norway*, February 11, 2019. As of August 13, 2020:
https://www.tv2.no/a/10406767/

"Peskov: Rossiya ne mozhet byt' v storone, kogda russkim grozyat nasiliem," *RIA Novosti*, March 7, 2014.

Pettyjohn, Stacie L., and Becca Wasser, *Competing in the Gray Zone: Russian Tactics and Western Responses*, Santa Monica, Calif.: RAND Corporation, RR-2791-A, 2019. As of October 31, 2020:
https://www.rand.org/pubs/research_reports/RR2791.html

Pirseyedi, Bobi, *The Small Arms Problem in Central Asia: Features and Implications*, Geneva: United Nations Institute for Disarmament Research, 2000.

Plokhy, Serhii, and M. E. Sarotte, "The Shoals of Ukraine: Where American Illusions and Great-Power Politics Collide," *Foreign Affairs*, January/February 2020.

Political Economy Research Institute, *Modern Conflicts: Conflict Profile, Tajikistan (1992–1998)*, Amherst, Mass.: University of Massachusetts Amherst, undated.

Polity IV Project, "Polity IV Individual Country Regime Trends, 1946–2013," webpage, June 6, 2014. As of September 1, 2020:
https://www.systemicpeace.org/polity/polity4.htm

President of Russia, "Press Conference for the Russian and Foreign Media," speech, Moscow, January 31, 2006. As of September 1, 2020:
http://www.kremlin.ru/eng/speeches/2006/01/31/0953_
type82915type82917_100901.shtml

———, "Speech and the Following Discussion at the Munich Conference on Security Policy," transcript, Munich, February 10, 2007. As of August 30, 2020:
http://en.kremlin.ru/events/president/transcripts/24034

———, "Press Statement and Answers to Journalists' Questions Following a Meeting of the Russia-NATO Council," transcript, April 4, 2008a. As of September 1, 2020:
http://en.kremlin.ru/events/president/transcripts/24903

———, "Interview Given by Dmitry Medvedev to Television Channels Channel One, Rossia, NTV," webpage, August 31, 2008b. As of October 31, 2020:
http://en.kremlin.ru/events/president/transcripts/48301

———, "Address by the President of the Russian Federation," speech, Moscow: The Kremlin, March 18, 2014a. As of September 1, 2020:
http://en.kremlin.ru/events/president/news/20603

———, "Transcript: Putin Says Russia Will Protect the Rights of Russians Abroad," *Washington Post*, March 18, 2014b.

———, "Meeting of the Valdai International Discussion Club," excerpts of transcript, Sochi, October 24, 2014c. As of October 31, 2020:
http://en.kremlin.ru/events/president/news/46860

———, "70th Session of the UN General Assembly," transcript, September 28, 2015a. As of October 31, 2020:
http://www.en.kremlin.ru/events/president/transcripts/statements/50385

———, "Meeting with Government Members," transcript, Novo-Ogaryovo, Moscow Region, September 30, 2015b. As of October 31, 2020:
http://en.kremlin.ru/events/president/news/50401

———, "Meeting of the Valdai International Discussion Club," Sochi, October 18, 2018. As of October 31, 2020:
http://en.kremlin.ru/events/president/news/58848

Pujol, Catherine, "Some Reflections on Russian Involvement in the Tajik Conflict, 1992–1993 (with Chronology)," in Mohammad-Reza Djalili, Frédéric Grare, and Shirin Akiner, eds., *Tajikistan: The Trials of Independence*, New York: Routledge, 1998, pp. 99–118.

"Putin Claims Russia Was 'Forced to Defend Russian-Speaking Population in Donbass,'" *The Interpreter*, October 12, 2016.

"Putin schitaet nedopustimym povtorenie v Sirrii 'liviiskogo' stsenariya," *RIA Novosti*, June 2, 2017.

Rasler, Karen A., and William R. Thompson, "Contested Territory, Strategic Rivalries, and Conflict Escalation," *International Studies Quarterly*, Vol. 50, No. 1, March 2006, pp. 145–167.

Ray, James L., "Friends as Foes: International Conflict and Wars Between Formal Allies," in Charles S. Gochman and Alan Ned Sabrosky, eds., *Prisoners of War? Nation-States in the Modern Era*, Lexington, Mass.: Lexington Books, May 1, 1990, pp. 73–91.

Recknagel, Charles, "A Side-by-Side Comparison of the Russian and Ukrainian Militaries," *The Atlantic*, March 19, 2014.

Reed, William, "A Unified Statistical Model of Conflict Onset and Escalation," *American Journal of Political Science*, Vol. 44, No. 1, January 2000, pp. 84–93.

Rempfer, Kyle, "U.S. Forces Concluding Relief Efforts in Mozambique," *Air Force Times*, April 12, 2019.

Renshon, Jonathan, "Status Deficits and War," *International Organization*, Vol. 70, No. 3, 2016, pp. 513–550.

Renz, Bettina, *Russia's Military Revival*, Cambridge, United Kingdom: Polity, 2018.

"Reports: Georgia Returns Seized Cargo to Russian Peacekeepers," *Civil Georgia*, August 15, 2005.

"Reports: Georgian Airspace Violated Again," *Civil Georgia*, November 16, 2004.

Robinson, Linda, Todd C. Helmus, Raphael S. Cohen, Alireza Nader, Andrew Radin, Madeline Magnuson, and Katya Migacheva, *Modern Political Warfare: Current Practices and Possible Responses*, Santa Monica, Calif.: RAND Corporation, RR-1772-A, 2018. As of October 31, 2020: https://www.rand.org/pubs/research_reports/RR1772.html

Rondeaux, Candace, *Decoding the Wagner Group: Analyzing the Role of Private Military Security Contractors in Russian Proxy Warfare*, Washington, D.C.: New America, November 7, 2019.

Roth, Andrew, "Central African Republic Considers Hosting Russian Military Base," *The Guardian*, October 25, 2019.

Rubin, Barnett R., "The Fragmentation of Tajikistan," *Survival*, Vol. 35, No. 4, 1993, pp. 71–91.

"Russia Denies Violating Georgia's Airspace," *Civil Georgia*, November 11, 2004.

"Russia Recognizes Abkhazia, South Ossetia," *Radio Free Europe/Radio Liberty Newsline*, August 26, 2008.

"Russia Refuses Deploying Weapons in Georgia's Breakaway Region," *Civil Georgia*, February 7, 2003.

"Russia Slams Georgia over South Ossetia," *Civil Georgia*, October 3, 2005.

"Russian Border Guards' Presence on Border with Turkey Important for Armenia—Premier," *TASS*, July 26, 2018.

"Russian Military Personnel in Mozambique: Bringing Another African Nation Closer to Moscow," Warsaw Institute, September 30, 2019.

"Russian Peacekeeper Kidnapped in Georgia Freed," Agence France Presse, October 1, 2003.

"Russian Peacekeeper Kidnapped in Georgia: Official," Agence France Presse, September 28, 2003.

"Russian Peacekeepers Accused of Smuggling in Abkhazia, S. Ossetia," *Civil Georgia*, August 12, 2005.

"Russian Peacekeepers Search for Kidnapped Solider," *Civil Georgia*, September 30, 2003.

"Russian Ship Flees Odessa," *Washington Post*, April 10, 1994.

"Russian Spy Suspects Sentenced to Pre-Trial Detention," *Civil Georgia*, September 29, 2006.

"Saakashvili: Certain Forces in Russia Prepare for Aggression Against Georgia," *Civil Georgia*, July 11, 2004.

"Saakashvili Vows to Hunt Down 'Spies' in Abkhazia, S. Ossetia," *Civil Georgia*, October 2, 2006.

"Sailors Mutiny and Take a Ship to Ukraine," *New York Times*, July 22, 1992.

<قلقون الله: قلقون من تواجد مرتزقة روس في حقل الشرارة> [Sanalla: We Are Unsettled by the Presence of Russian Mercenaries in the al-Sharara Field], *Libya al-Ahrar*, June 24, 2020.

Sanger, David E., *Confront and Conceal: Obama's Secret Wars and Surprising Use of American Power*, New York: Crown Publishers, 2012.

———, "Hack of Saudi Petrochemical Plant Was Coordinated from Russian Institute," *New York Times*, October 23, 2018.

Sauer, Pjotr, "7 Kremlin-Linked Mercenaries Killed in Mozambique in October—Military Sources," *Moscow Times*, October 31, 2019.

Saul, Jonathan, "Exclusive: Russia Steps Up Military Lifeline to Syria's Assad—Sources," Reuters, January 17, 2014.

Schenker, David, "Assistant Secretary for Near Eastern Affairs David Schenker—Special Briefing," transcript, Washington, D.C.: Press Correspondent's Room, November 26, 2019.

Schwirtz, Michael, and Gaelle Borgia, "How Russia Meddles Abroad for Profit: Cash, Trolls and a Cult Leader," *New York Times*, November 11, 2019.

Searcey, Dionne, "Gems, Warlords, and Mercenaries: Russia's Playbook in Central African Republic," *New York Times*, September 30, 2019.

Security Assistance Monitor, "Security Assistance Database," data files, undated. As of October 1, 2020:
https://securityassistance.org/security-sector-assistance/

Seiffert, Jeanette, "The Significance of the Donbas," *DW*, April 15, 2014.

Senese, Paul D., "Geographical Proximity and Issue Salience: Their Effects on the Escalation of Militarized Interstate Conflict," *Conflict Management and Peace Science*, Vol. 15, No. 2, September 1996, pp. 133–161.

"Senior Russian Senator Speaks of Georgia's 'Spy Mania,'" *Civil Georgia*, September 29, 2006.

Sepashvili, Giorgi, "Controversial Reports over Deployment of Extra Arms Flare Tensions in Ossetia," *Civil Georgia*, June 13, 2004.

Singer, J. David, "Accounting for International War: The State of the Discipline," *Journal of Peace Research*, Vol. 18, No. 1, 1981, pp. 1–18.

Singer, P. W., "Stuxnet and Its Hidden Lessons on the Ethics of Cyberweapons," *Case Western Reserve Journal of International Law*, Vol. 47, No. 1, 2015, pp. 79–86.

Smirnov, Mikhail, "Like a Sack of Potatoes: Who Transferred the Crimean Oblast to the Ukrainian SSR in 1952–54 and How It Was Done," *Russian Politics and Law*, Vol. 53, No. 2, 2015, pp. 32–46.

Smith, Alastair, "Testing Theories of Strategic Choice: The Example of Crisis Escalation," *American Journal of Political Science*, Vol. 43, No. 4, October 1999, pp. 1254–1283.

Smith, Matt, and Victoria Butenko, "Ukraine Says It Retakes Building Seized by Protesters," *CNN*, April 7, 2014.

Snegovaya, Maria. "What Factors Contribute to the Aggressive Foreign Policy of Russian Leaders?" *Problems of Post-Communism*, Vol. 67, No. 1, 2020, pp. 93–110.

Snow, Shawn, "AFRICOM Demands Return of U.S. Drone Shot Down by Russian Air Defenses over Libya," *Military Times*, December 10, 2019.

Sobczak, Blake, "The Inside Story of the World's Most Dangerous Malware," *E&E News*, March 7, 2019.

Soufan Group, *Foreign Fighters: An Updated Assessment of the Flow of Foreign Fighters into Syria and Iraq*, New York, December 2015.

Spearin, Christopher, "NATO, Russia and Private Military and Security Companies: Looking into a Dark Reflection," *RUSI Journal*, Vol. 163, No. 3, August 8, 2018.

Splidsboel-Hansen, Flemming, "The Outbreak and Settlement of Civil War: Neorealism and the Case of Tajikistan," *Civil Wars*, Vol. 2, No. 4, 1999, pp. 1–22.

Stewart, Phil, Idrees Ali, and Lin Noueihed, "Exclusive: Russia Appears to Deploy Forces in Egypt, Eyes on Libya Role—Sources," Reuters, March 13, 2017.

Stupachenko, Ivan, "Russia and Norway Clash over Status of Waters Around Spitsbergen/Svalbard," *SeafoodSource*, February 27, 2020.

"Sudan Investigating Transfer of Guards from UAE to Libyan Oil Port—Ministry," Reuters, January 28, 2020.

Sukhankin, Sergey, "War, Business and 'Hybrid' Warfare: The Case of the Wagner Private Military Company (Part Two)," *Eurasia Daily Monitor*, Vol. 15, No. 61, April 23, 2018a.

———, "'Continuing War by Other Means': The Case of Wagner, Russia's Premier Private Military Company in the Middle East," Jamestown Foundation, July 13, 2018b.

———, "The 'Hybrid' Role of Russian Mercenaries, PMCs and Irregulars in Moscow's Scramble for Africa," Jamestown Foundation, January 10, 2020a.

———, "Russian Mercenaries Pour into Africa and Suffer More Losses (Part One)," *Eurasia Daily Monitor*, Vol. 17, No. 6, January 21, 2020b.

———, "Continuation of Policy by Other Means: Russian Private Military Contractors in the Libyan Civil War," *Terrorism Monitor*, Vol. 18, No. 3, February 7, 2020c.

———, "Russian PMCs and Irregulars: Past Battles and New Endeavors," Jamestown Foundation, May 13, 2020d.

Taylor, Ann, "Days of Protest in Ukraine," *The Atlantic*, December 2, 2013.

"Tbilisi Wants Moscow to Explain Incident in Abkhaz Conflict Zone," *Civil Georgia*, March 22, 2005.

Theohary, Catherine A., *Defense Primer: Information Operations*, Washington, D.C.: Congressional Research Service, IF10771, December 15, 2020.

Thomas, Timothy L., "Deterring Information Warfare: A New Strategic Challenge," *Parameters*, Vol. 26, No. 4, Winter 1996–97, pp. 81–91.

———, *Russian Military Thought: Concepts and Elements*, McLean Va.: MITRE Corporation, MP190451V1, August 2019.

Tigner, Brooks, "Electronic Jamming Between Russia and NATO Is Par for the Course in the Future, but It Has Its Risky Limits," Atlantic Council, November 15, 2018.

Tisdall, Simon, and Rory Carroll, "Russia Sets Terms for Ukraine Deal as 40,000 Troops Mass on Border," *The Guardian*, March 30, 2014.

Toal, Gerard, *Near Abroad: Putin, the West, and the Contest over Ukraine and the Caucasus*, New York: Oxford University Press, 2017.

Townsend, Stephen J., "Statement of General Stephen J. Townsend, United States Army Commander, United States Africa Command, Before the Senate Armed Services Committee," Washington, D.C., January 30, 2020.

Traynor, Ian, and Oksana Grytsenko, "Ukraine Suspends Talks on EU Trade Pact as Putin Wins Tug of War," *The Guardian*, November 21, 2013.

Trenin, Dmitri, "Russia's Spheres of *Interest*, Not *Influence*," *Washington Quarterly*, Vol. 32, No. 4, October 2009, pp. 3–22.

Troianovski, Anton, and Ellen Nakashima, "How Russia's Military Intelligence Agency Became the Covert Muscle in Putin's Duels with the West," *Washington Post*, December 28, 2018.

"Troops March in Tskinvali Marking 'Independence,'" *Civil Georgia*, September 20, 2005.

Tsygankov, Andrei P., "Vladimir Putin's Last Stand: The Sources of Russia's Ukraine Policy," *Post-Soviet Affairs*, Vol. 31, No. 4, February 2015, pp. 279–303.

Tuathail, Gearóid Ó, "Russia's Kosovo: A Critical Geopolitics of the August 2008 War over South Ossetia," *Eurasian Geography and Economics*, Vol. 49, No. 6, 2008, pp. 670–705.

Tucker, Patrick, "Mozambique Is Emerging as the Next Islamic Extremist Hotspot," *Defense One*, July 6, 2020.

"Tuvalu Retracts Abkhazia, S. Ossetia Recognition," Civil Georgia, March 31, 2014.

"Two Helicopters Violate Georgian Airspace," *Civil Georgia*, October 28, 2004.

"Ukraine Crisis: Casualties in Sloviansk Gun Battles," *BBC News*, April 13, 2014.

"Ukraine Protests After Yanukovych EU Deal Rejection," *BBC News*, November 30, 2013.

"Ukraine Takes Action in Slovyansk; West Condemns Russian Involvement," Radio Free Europe/Radio Liberty, April 13, 2014.

"Ukrainian Clashes with Pro-Russian Separatists Turn Deadly," *Washington Post*, April 13, 2014.

U.S. Africa Command Public Affairs, "Declining Security in Libya Results in Personnel Relocation, Agility Emphasis," press release, Stuttgart, Germany, April 7, 2019.

U.S. Department of Defense, *Department of Defense: Strategy for Operations in the Information Environment*, Washington, D.C., June 2016.

———, *Joint Concept for Operating in the Information Environment (JCOIE)*, Washington, D.C., July 25, 2018.

———, *DoD Dictionary of Military and Associated Terms*, Washington, D.C., January 2021.

U.S. Department of the Treasury, "Treasury Designates Individuals and Entities Involved in Ongoing Conflict in Ukraine," press release, Washington, D.C., June 20, 2017.

Vasquez, John A., "Distinguishing Rivals That Go to War from Those That Do Not: A Quantitative Comparative Case Study of the Two Paths to War," *International Studies Quarterly*, Vol. 40, No. 4, December 1996, pp. 531–558.

———, *The War Puzzle Revisited*, Cambridge, United Kingdom: Cambridge University Press, 2009.

Venter, Al J., "A Dirty Little War in Mozambique," *Key Aero*, No. 386, May 2020, pp. 74–79.

Vest, Nathan, and Colin P. Clarke, "Is the Conflict in Libya a Preview of the Future of Warfare?" *Defense One*, June 2, 2020.

Volz, Dustin, "Researchers Link Cyberattack on Saudi Petrochemical Plant to Russia," *Wall Street Journal*, October 23, 2018.

Walker, Shaun, "New Evidence Emerges of Russian Role in Ukraine Conflict," *The Guardian*, August 18, 2019.

Waltz, Kenneth N., *Theory of International Politics*, New York: Random House, 1979.

Ward, Michael D., Brian D. Greenhill, and Kristin M. Bakke, "The Perils of Policy by P-Value: Predicting Civil Conflicts," *Journal of Peace Research*, Vol. 47, No. 4, 2010, pp. 363–375.

Wehrey, Frederic, "With the Help of Russian Fighters, Libya's Haftar Could Take Tripoli," *Foreign Policy*, December 5, 2019.

Weiss, Jessica Chen, *Powerful Patriots: Nationalist Protest in China's Foreign Relations*, New York: Oxford University Press, 2014.

Wiegand, Krista E., "Militarized Territorial Disputes: States' Attempts to Transfer Reputation for Resolve," *Journal of Peace Research*, Vol. 48, No. 1, January 2011, pp. 101–113.

Wiener-Bronner, Danielle, "Another Ukrainian City Wants Its Independence," *The Atlantic*, April 7, 2014.

Wilson, Scott, "Obama Dismisses Russia as 'Regional Power' Acting out of Weakness," *Washington Post*, March 25, 2014.

Wilson, Steve, Peter Foster, and Katie Grant, "Ukraine as It Happened: Urgent Calls for Calm as West Faces Biggest Confrontation with Russia Since Cold War," *The Telegraph*, March 2, 2014.

Wojnowski, Zbigniew, "Economic Tensions Worsen Unrest in Eastern Ukraine," *Al Jazeera America*, March 25, 2014.

World Bank, "World Bank Open Data," webpage, undated-a. As of October 2, 2020:
https://data.worldbank.org/

———, World Development Indicators, data catalog, undated-b. As of October 2, 2020:
https://datacatalog.worldbank.org/dataset/world-development-indicators

———, "World Integrated Trade Solution," webpage, undated-c. As of October 2, 2020:
https://wits.worldbank.org

Yapparova, Liliya, "A Small Price to Pay for Tripoli," *Meduza*, October 2, 2019.

Yekelchyk, Serhy, *The Conflict in Ukraine: What Everyone Needs to Know*, New York: Oxford University Press, 2015.

Yuhas, Alan, and Raya Jalabi, "Ukraine Crisis: Why Russia Sees Crimea as Its Naval Stronghold," *The Guardian*, March 7, 2014.

Yusin, Maksim, and Sergei Strokan', "Ni mira, ni Pal'mira," *Kommersant*, May 22, 2015.

Zaborsky, Victor, *Crimea and the Black Sea Fleet in Russian-Ukrainian Relations*, Cambridge, Mass.: Harvard University, Kennedy School of Government, CSIA Discussion Paper 95-11, September 1995.

Zalewski, Piotr, "Russian Separatism Gains Ground in Eastern Ukraine," *TIME*, March 19, 2014.

Zevelev, Igor A., *Russia and Its New Diasporas*, Washington, D.C.: United States Institute of Peace, February 2001.

———, *Russian National Identity and Foreign Policy*, Washington, D.C.: Center for Strategic and International Studies, Russia and Eurasia Program, December 2016.

"Zourabichvili: Ganmukhuri Incident Could Be a Provocation," *Civil Georgia*, March 22, 2005.

Milton Keynes UK
Ingram Content Group UK Ltd.
UKHW020947071123
432119UK00009B/57

9 781977 407948